Workers and the World

Lorenzo Feltrin is a researcher at Ca' Foscari University of Venice. He grew up in Treviso, close to Venice and its industrial hub, Porto Marghera. In Treviso, he took part in the occupations that established the Django Social Centre. He is active in international social movement networks mobilising on labour and environmental issues, such as the ex-GKN Florence autoworkers' struggle for a just transition.

Workers and the World

Fighting Ecological Crisis from Within

Lorenzo Feltrin

London • New York

First published by Verso 2026

Cover image: from *La unidad: Órgano oficial de los obreros de ENAMI – Las Ventanas*, no. 19 (May–June 1971).

The manufacturer's authorized representative in the EU for product safety (GPSR) is LOGOS EUROPE, 9 rue Nicolas Poussin, 17000, La Rochelle, France contact@logoseurope.eu

1 3 5 7 9 10 8 6 4 2

Verso
UK: 6 Meard Street, London W1F 0EG
US: 207 East 32nd Street, New York, NY 10016
versobooks.com

Verso is the imprint of New Left Books

ISBN-13: 978-1-80429-782-7
ISBN-13: 978-1-80429-783-4 (UK EBK)
ISBN-13: 978-1-80429-784-1 (US EBK)

British Library Cataloguing in Publication Data
A catalogue record for this book is available from the British Library

Library of Congress Cataloging-in-Publication Data
A catalog record for this book is available from the Library of Congress

Typeset in Minion by Hewer Text UK Ltd, Edinburgh
Printed and bound by CPI Group (UK) Ltd, Croydon CR0 4YY

I lay down a few instants more
in the warehouse of these vivid stories
to retain by force onto the iris
the echo of accidental clouds
rolling over Occidental anthills and laughing
at Pacific Oceans that seem Indian ink
at myself, at a compromised celestial body
and I write . . .
These abandoned stories
like the building sites at the edges of neighbourhoods.
We grew up disorderly, like these turbid peripheries
of which I hazard a paraphrasis.
Between caresses icier than a 'Maybe',
in the asthma displays,
scarcities and respective anaesthetics.
Before going out I take the last snapshot
of my room, what chaos.
Rust, sheets and leaves everywhere, rough syllables, dew
residues of us on coffee grounds,
the road leading to Porto Marghera.
Let the wind carry everything away, I leave the windows open
and these parentheses closed, together with the squares
with low pressure and grey expression
evacuated with white spirit hydrants.
I put in the suitcase the translation in wrong rhymes and dirty notes
of this anthology of anti-stories and I leave,
aware that on the upper floors they're hounding the algorithm of smiles
to photocopy us while every day we sell them our shade
for a comfortable spot among the cogs.
I write a storm so that the sky may fall upon their heads . . .

Alberto 'Dubito' Feltrin, 'Storie abbandonate', *Santa Bronx*

Contents

Preface

In 2008, as a twenty-year-old student in philosophy, I met oral history practitioner and countercultural agitator Marco Philopat, an encounter that generated an enduring friendship and, in a way, changed my life. Philopat gave me a crash course on how to conduct biographical interviews and encouraged me to push forward with a book idea on the grime and dubstep scenes in London, where I had just spent a few months working in Notting Hill's Marks & Spencer. I thus returned to the British capital for a summer, generously hosted in a Hackney squat, working part-time at the Euston Station Burger King, and interviewing DJs, producers, MCs, activists and scholars. The result was, no doubt, immature, but it made me realise I was more gifted as a sociologist than as a music critic or philosopher.[1] In the summer of 2011, moved by a youthful and naïf enthusiasm for the unfolding 'Arab Spring', I spent a month in Cairo (hosted through Couch Surfing), interviewing kids of my age who were taking part in the Tahrir Square occupation.[2] It was the beginning of a long journey.

1 Lorenzo Feltrin, *Londra zero zero*, Milan: Agenzia X, 2010.

2 Lorenzo Feltrin, 'La libertà è tutto: Racconti della Rivoluzione Egiziana', *GlobalProject*, 29 August 2011, globalproject.info.

Workers and the World: Fighting Ecological Crisis from Within presents the results of research spanning a period of ten years. Between 2014 and 2018, I was a PhD student in the Politics and International Studies (PAIS) department of the University of Warwick, under the supervision of Nicola Pratt. Following that, I worked as a research assistant in Warwick's Sociology Department, for the project 'Toxic Expertise: Environmental Justice and the Global Petrochemical Industry', led by Alice Mah. Finally, between 2021 and 2024, I was a Leverhulme Early Career Fellow at the University of Birmingham, mentored by David Bailey. The funding for this research came from a combination of the University of Warwick Chancellor's Scholarship, the European Research Council under the European Union's Horizon 2020 research and innovation programme (Grant Agreement No. 639583) and the Leverhulme Trust (Early Career Fellowship 2020-004). In addition to being affiliated with the University of Warwick and the University of Birmingham, I have taken up Visiting Fellowships at the University of Padua and the University of Chile.

The book incorporates fieldwork from four cases, including more than 120 semi-structured interviews with industrial workers, residents of communities adjacent to the industrial sites, trade unionists, activists and experts, as well as extensive archival research and analyses of socio-economic statistics, labour legislation, memoirs and other documents produced by trade unions, social movement organisations and public agencies. Some of the book's empirical materials have already been released in different forms, but they have been reworked here through a comprehensive theoretical framework.

That said, the product of research is not only the text submitted once the writing is over. There is also the very subjectivity of the researcher. Here, the 'authors' are all those I have been in touch with at different times during this period. The task of getting by would have been impossible without them, and I hope that I have been able to help others in turn. The overall result is that this book is something I would not have written if I could now go back in time. This is because I have changed in the process, and would carry out the research differently were I to undertake it today.

Firstly, it would be untrue to claim that the four cases were selected according to a carefully planned research design. There was a strong element of chance in the assemblage of this unlikely quartet, related to my shifting personal circumstances and the fact that the book brings together research from different fixed-term jobs. More fundamentally, my awareness of the dangers of 'academic extractivism' started from a fairly low point.[3] I have since changed my practices over time, setting up institutional links with local universities, releasing open access research outputs in languages the participants can read, and planning restitution arrangements with them. Nonetheless, carrying out fieldwork in four different and faraway places makes it difficult to maintain regular and durable relations with all communities of participants. I thus apologise for these shortcomings.

Ultimately, I agree with critiques of the incumbent hierarchies of global knowledge production.[4] While these epistemological hierarchies are unlikely to disappear for as long as steep economic inequalities resulting from the international division of labour persist, the two dialectically reinforce each other and must therefore be addressed together. The question is how to tackle them. As with most political questions, individual withdrawal is unlikely to change anything. Therefore, I have not concluded that one's research should be restricted to places close to one's background. Indeed, on the obverse side of this, it is important that researchers from the Global South be able to do more research on the Global North than is currently the case. Academics and research institutions in the Global North are not exempt from contributing to transforming the existing regime of knowledge production. This should point towards a rebalancing of who gets funding (and visas!) and steer considerations on how that funding should be used, as part of a broader collective struggle to put knowledge at the service of a more egalitarian world.

3 Melany Cruz and Darcy Luke, 'Methodology and Academic Extractivism: The Neo-Colonialism of the British University', *Third World Thematics* 5(1–2), 2020, 154–70.

4 For example, Syed F. Alatas, 'Academic Dependency and the Global Division of Labour in the Social Sciences', *Current Sociology* 51(6), 2003, 599–613.

I would like to thank all the participants in the research, who generously gave up their time and attention to talk to me about their struggles and hopes. Some of them are no longer with us, including the Tunisian militants Mohamed Ali Arous, Salah Hamzaoui, Youssef Salhi and Salah Zeghidi, Chilean docker and union representative Fabian Ordenes Silva, and Italian teacher and community volunteer Luisa Colio. I am grateful for having had the chance to meet them and would like to honour their memory here. Professor Salah Hamzaoui had written extensively about Tunisian trade unionism, and I shared with him many cherished discussions beyond the interview itself.[5]

This book has benefited from the intellectual exchanges I have had with many colleagues, not least: Rutvica Andrijasevic, Martín Arboleda, Giulia Arrighetti, Maurizio Atzeni, Mustapha Azaitraoui, Dario Azzellini, Francesco Bagnardi, Stefano Barone, Sarah Barrières, Karen Bell, Monji Ben Chaaben, Oscar Berglund, Charlotte Sophia Bez, Koenraad Bogaert, Irene Bono, David Brown, Pietro Brunello, Filipe Calvão, Clara Capelli, Alessia Carnevale, John Chalcraft, Riccardo Emilio Chesta, Lorenzo Cini, Gabriella Cioce, Thom Davies, Damiano De Facci, Gianni Del Panta, Elio Di Muccio, Pietro Di Paola, Ana Cecilia Dinerstein, Moutaa Amin El Waer, Fátima Fernandez, Antonio Ferraz De Oliveira, Patricio Flores Silva, Leopoldina Fortunati, Emma Foster, Francesca Gabbriellini, Franco Galdini, Manuela Galetto, Davide Gallo Lassere, Ferruccio Gambino, Stefano Gasparri, Manoel Gehrke, Lorenzo Genito, Craig Gent, Gennaro Gervasio, Katharina Grueneisl, Moncef Guebsi, Hamza Hamouchene, David Harvie, Choukri Hmed, Paola Imperatore, Calvin Jephcote, Tereza Jermanová, Peter Kerr, Emanuele Leonardi, Simon Lobach, Anna Lora-Wainwright, Chiara Loschi, Loretta Lou, Vincenzo Macarrone, Alice Mah, Marco Marrone, Lydia Medland, Alessandra Mezzadri, Alexis Moraitis, Javier Moreno Zacares, Aya Nassar, Grettel Navas, Cecilia Pasini, Luigi Pellizzoni, Pablo Pérez Ahumada, Daniela Pioppi, Iain Pirie, Giorgio Pirina, Piermarco Piu, Stefano Pontiggia, Nicola Quondamatteo, Al Amin Rabby, Patricia

5 Salah Hamzaoui, 'Pratiques syndicales et pouvoir politique: Pour une sociologie des cadres syndicaux (cas de la Tunisie)', PhD thesis, Université Paris 7, 2013.

Daniela Retamal, Alessandro Rivera Magos, Diane Robert, Juan Pablo Rodríguez, Devi Sacchetto, Mariam Salehi, Ester Sigillò, Živa Šketa, Harshal Sonekar, Arianna Tassinari, Magdalena Taube, Lisa Tilley, Francesco Vacchiano, Thomas Verbeek, Lorenzo Vianelli, Chris Waite, Jamie Woodcock, Krystian Woznicki, Gökçe Yeniev, Fayrouz Yousfi, Hèla Yousfi and Gilda Zazzara. Special gratitude goes to David Bailey, Will Berrington, Jack Copley and Christine Emmett, who have provided feedback on previous drafts of the book.

It is impossible to mention even a fraction of the friends I would like to acknowledge. I will settle for a collective thanks to the community of Treviso's Django Social Centre, a former military complex in my hometown turned into a space for political, cultural and mutual aid activities. I had known many of these comrades long before we started occupying abandoned buildings together, joining the picket lines of logistics workers, blocking house evictions, organising climate and women's strikes, and much more. In particular, I would like to express my gratitude to those who have devoted larger chunks of their lives to this project, which brought real change to a town that was known nationally as a fascist stronghold. I am also thankful to the editorial board of the social movement website globalproject.info, which has regularly hosted my contributions.

I owe endless gratitude to my parents, Paolo Feltrin and Renata Gherlenda, for their love, their support and their understanding. At the risk of eliciting national stereotypes, I also include the 'extended family': my aunts, uncles, cousins and grandparents. A loving thanks to Gabriela Julio Medel for the countless precious things we have shared in our journey.

My brother, Alberto 'Dubito' Feltrin, left us in 2012, aged only twenty. This book is dedicated to him. Alberto had a precocious talent for poetry, music and the visual arts. He had developed an imaginary about the 'rusted peripheries' of Italy's Northeast, with a deeply human, social and political sensitivity. Together, we drew paths in the cracks of ageing industrial areas, dreamed about metropoles while walking along countryside railway tracks, and hypothesised apocalypses from our neighbourhood's playground only to brush them off with a laugh. With a pinch of fantasy, Treviso's condos and wheat fields became urban jungles and open prairies. We liked going to Marghera – maybe that is what brought me back eventually. On the trail of trans-provincial commodity chains, we were convinced that exploring deserted hangars along the canals was the key to figuring out how we got to become us, especially after the disorientation wreaked by the 2007 financial meltdown. Not having understood Alberto properly is my biggest regret. I have become an 'adult' alone, but there will always remain that part of me which used to jump over fences with Alberto at the edge of towns, and still stays with him in the dim glow of the rusted peripheries.

Introduction
Workers and the Ecological Crisis

On 24 August 2018, hundreds of protesters broke into the copper processing complex in Ventanas, Chile, smashing windows with stones, to the dismay of its astonished employees. The crowd, composed mostly of local residents, was enraged about the harsh industrial toxicity afflicting the area. Fifteen years earlier, in Venice, Italy, a march of Porto Marghera's petrochemical workers nearly clashed with a sit-in held by environmental community groups demanding an end to hazardous production in the vicinity of their neighbourhoods. Only frantic negotiations kept the two groups apart. These are just two out of countless examples of tensions between workplace and community organisations over noxious productive facilities. To the extent that the communities living near such installations tend to be disproportionately working class, we can largely view these localised incidents as conflicts internal to the class that must live by its labour. However, they are also part of a much larger battlefield, shaped – as we will see – by the deeper antagonism between total labour and capital.

We are *in* the ecological crisis, and not just as victims of an environmental devastation that is unequally distributed along intersecting hierarchies of class, 'race' and gender on a global scale. We are part of the crisis because in our societies the vast majority of us rely on capitalist work to pay for the things we need to survive. Our dependence on capitalist work means we also depend on the infinite growth of commodity production that defines capitalism and drives the ecological crisis. When this expansion breaks down, as it does in times of economic crisis, unemployment rises and livelihoods become more precarious. The reproduction of the working class as a class is tied to the reproduction of capital as a social relation, and vice versa. As Karl Marx wrote: 'The reproduction of labour-power . . . which cannot get free of capital . . . forms, in fact, a factor in the reproduction of capital itself.'[1] For our purposes, one might add that the capitalist reproduction of labour power is also an essential factor in the reproduction of the ecological crisis.

Because they are dispossessed of the means of production (land, machinery, inputs, etc.), workers are forced to enter the labour process as labour power in order to make a living. They thus become a part of capital, which Marx called 'variable capital' because it produces additional capital, thus reproducing capitalist society. To the extent that – as will be shown – capitalist production is inherently noxious, *capital is ecological crisis*. Hence workers as the living, productive articulation of capital are an articulation of the ecological crisis and must constantly deepen it if they are to survive. Such 'job blackmail' – the forced choice between contributing to environmental degradation or losing one's income – does not only concern highly noxious productive sites; it is, rather, an intrinsic and transversal feature of capitalism, which becomes manifest with variable intensity in different contexts.

Nonetheless, workers' insertion in capital accumulation also has an antagonistic face. This is rooted in the workers' very separation from the means of production, which is the condition of fully developed capitalism. As a result of their dispossession, workers must satisfy many of their needs through the market. Under impersonal competition, the wealth produced within capitalist social relations becomes commodified and regulated by value, the average labour time needed to produce all kinds of goods and services. The law of value is the constant pressure for

1 Karl Marx, *Capital: A Critique of Political Economy*, vol. 1, London: Penguin, 1976 [1867], 763–4.

commodity prices to ultimately reflect this social average, a pressure that qualitatively moulds production itself. Workers thus experience market discipline as an external force imposed on them by capitalists and managers (who are just as dependent on capitalist work, albeit on the capitalist work of others, which they own, organise or supervise under the constraints of the same market discipline). This creates a structural antagonism between the needs of the working class and those of capital, at times self-conscious and open, at times unacknowledged and subterranean. Such an antagonism constitutes a limit to the imperatives of capital accumulation, and thus a potential force *against* the ecological crisis. The ultimate horizon of this antagonism is the full self-negation of the working class through a movement beyond capitalism and towards a classless mode of production.

Contrary to productivist versions of Marxism, which see work and technology in capitalism as progressive forces to be liberated from the fetters of capitalist social relations, the Italian New Left current known as *operaismo* saw capitalist work and technology as qualitatively shaped by such social relations. Hence, the provocative advocacy of the refusal of capitalist work itself. Not liberation of work from capitalism, but *liberation from capitalist work*. In the words of *operaista* author Mario Tronti: 'A working-class struggle against work, struggle of the worker against himself [*sic*] as worker, labour-power's refusal to become labour.'[2] On these bases, Tronti emphasised the dual nature of workers in capitalism as both a factor of capital (labour power) and as a force antagonistic to it (the working class):

> When we are talking about the working class within the system of capital, the same productive force really can be counted twice: one time as a force that *produces* capital and another time as a force that *refuses* to produce it; one time *in* capital, another time *against* capital.[3]

2 Mario Tronti, *Workers and Capital*, London: Verso, 2019 [1966], 401. Some translations were edited by the author.

3 Ibid., 271.

Yet, as inspiring as this might sound, within today's environmental conflicts, the duality of the proletarian condition *appears* to be split across different segments of the working class. On the one hand, the workers employed in highly noxious facilities struggle *for their own work*, often relinquishing any ambitions of radically transforming the labour process and the social relations that shape it, prioritising instead the defence of jobs against closure threats. On the other hand, under-employed fenceline communities (that is, communities adjacent to such polluting industries) struggle *against the work of others*, with the closure of noxious productive sites being one of their key demands (even if individuals are sometimes happy to accept posts in such workplaces if offered).

This is why being against the ecological crisis is a challenge. It is the challenge of breaking the 'job blackmail' imposed on workers by creating multi-scalar convergences between workplace and community struggles and pushing towards a decommodification of production and nature capable of delivering a more egalitarian, sustainable and meaningful system of wealth creation. However, because the working class is fragmented along widely different occupational and residential arrangements, which too often fuels divisions between trade unionism and community environmentalism, such a convergence is far from automatic. It is this 'problem of convergence' that is the crux of the book you are reading.

Environmental justice and labour

Workers and the World explores the relationship between workers and capitalist noxiousness through a class analysis of conflicts around polluting productive facilities. The book is informed by empirical research presented in four case studies: two in the Global South – gas extraction in Kerkennah, Tunisia, and copper processing in Ventanas, Chile – and two in the Global North – petrochemical production in Grangemouth, United Kingdom, and Porto Marghera, Italy. Diverging from important Marxist accounts of the ecological crisis that focus on the dynamics of *capital accumulation*, this work – drawing on the tradition of *operaista* class composition analysis – offers an interpretation of *working-class struggles* against capitalist noxiousness and is intended as a toolkit for strengthening them.[4]

In 1977, *operaista* intellectual Sergio Bologna wrote:

> Unfortunately, there are many Marxist comrades who see the hypothesis of biological destruction as millenarian, and they smirk when they hear anything that might hint at 'ecology'. Seveso and many other cases have instead reminded us that the era of biological destruction has already begun, our children were born and are growing up in such a time. A process is underway that represents something more horrible than fascism.[5]

More recently, as the ecological crisis has inescapably become a central theme of political debate and mobilisation, increasing attention has been paid to the relationship between class and the environment. Two bodies of literature have risen to this challenge: environmental justice studies and political ecology, on the one hand, and environmental labour studies, on the other. The former has mainly focused on community-centred disputes, while the latter has brought to the fore workplace-centred organising.[6] A more recent strand, known as 'working-class environmentalism', has

4 See John B. Foster, *Marx's Ecology: Materialism and Nature*, New York: Monthly Review Press, 2000, for the former; Harry Cleaver, 'The Inversion of Class Perspective in Marxian Theory: From Valorisation to Self-Valorisation', in Werner Bonefeld, Richard Gunn and Kosmas Psychopedis (eds), *Open Marxism*, vol. 2, London: Pluto Press, 1992, 106–44, for the latter.

5 Sergio Bologna, 'La tribù delle talpe', *Primo Maggio* 8, 1977, 3–18, 8. All sources in French, Italian and Spanish were translated by the author. The Seveso disaster, which occurred on 10 July 1976, was a major industrial accident involving a chemical factory located in the Italian town of Seveso.

6 See Julian Agyeman, David Schlosberg, Luke Craven and Caitlin Matthews, 'Trends and Directions in Environmental Justice: From Inequity to Everyday Life, Community, and Just Sustainabilities', *Annual Review of Environment and Resources* 41, 2016, 321–40, for the former; Paul Hampton, *Workers and Trade Unions for Climate Solidarity: Tackling Climate Change in a Neoliberal World*, Abingdon-on-Thames: Routledge, 2015, for the latter.

assigned a central role to the problem of convergence between workplace-centred and community-centred struggles.[7] This book is a contribution to this emerging approach.

The environmental justice paradigm originated in the United States in the 1980s. It concentrated on racism as a source of the unequal distribution of environmental burdens, in light of the disproportionate siting of toxic industries in the vicinity of racialised communities. Nevertheless, class was always part of environmental justice analysis, which has, from the beginning, mapped the intersections between race and class as a predictor of living close to noxious facilities. Steve Lerner, for example, documented how industrial pollution engenders 'sacrifice zones' in low-income communities whose health becomes dispensable.[8] In Latin America, Javier Auyero and Débora Swistun showed how the Flammable community in Greater Buenos Aires was held captive by both industrial noxiousness and economic deprivation through 'toxic uncertainty'.[9]

Similarly, the 'environmentalism of the poor' frame pioneered by political ecologists Ramachandra Guha and Joan Martínez-Alier demonstrated how economically marginalised communities tend to be those most affected by environmental degradation and therefore have an interest in mobilising to protect their ecologies.[10] Héctor Alimonda combined political ecology with decolonial theory to mould a Latin American perspective of *ecologismo popular*, which usefully views the ecological crisis as rooted in the colonial conquest of nature that was a condition for the rise of capitalist modernity.[11] Ecofeminism has further explored how capitalism, coloniality

7 See, e.g., Stefania Barca, *Workers of the Earth: Labour, Ecology and Reproduction in the Age of Climate Change*, London: Pluto Press, 2024; Karen Bell, *Working-Class Environmentalism: An Agenda for a Fair and Just Transition to Sustainability*, London: Palgrave Macmillan, 2019.

8 Steve Lerner, *Sacrifice Zones: The Front Lines of Toxic Chemical Exposure in the United States*, Cambridge, MA: MIT Press, 2012.

9 Javier Auyero and Débora A. Swistun, *Flammable: Environmental Suffering in an Argentine Shantytown*, Oxford: Oxford University Press, 2008.

10 Ramachandra Guha and Joan Martínez-Alier, 'El ecologismo de los pobres', *Ecología Política* 8, 1994, 137–51.

11 Héctor Alimonda (ed.), *La naturaleza colonizada: Ecología política y minería en América Latina*, Buenos Aires: CLACSO, 2011.

and the patriarchy constitute a mutually reinforcing totality in which the racialised and gendered segments of the working class are disproportionally exposed to environmental degradation.[12]

Sometimes, environmental justice studies and political ecology appear to only see fenceline communities on one side and polluting companies on the other side. By treating polluting companies as an undifferentiated whole, such a simplified understanding obscures the labour–capital antagonism within noxious industries. However, at their best, environmental justice studies and political ecology do present a more refined analysis of environmental conflicts. For example, Robert Bullard's seminal work explicitly notes how 'the inherent conflict between the interest of capital and of labor' is at the root of socioecological hierarchies among different working-class segments.[13] Nonetheless, many empirical studies in this tradition restrict the scope of their analysis to community–industry conflicts, using a gradational conception of class centred on income levels and their residential distribution across territories. While these findings are crucial to making the case that the communities most exposed to environmental degradation are indeed disproportionally working class, a one-sided focus on community-centred grievances risks ignoring the workers employed by polluting companies. A full exploration of such workers' actual and potential role is strategically necessary, since this working-class segment is endowed with the structural power to halt or modify large-scale production from within.

Environmental labour studies, in many respects, provides a corrective to this shortcoming, with many analyses focusing on workplace-centred organising against environmental degradation.[14] Indeed, if a working-class 'proto-environmentalism' can be seen at the very origins of capitalism in

12 See, e.g., Maria Mies and Vandana Shiva, *Ecofeminism*, London: Zed Books, 1993; Melissa Moreano Venegas, Miriam Lang and Gabriela Ruales Jurado, 'Perspectivas de justicia climática desde los feminismos latinoamericanos y otros sures', rosalux.org, 2021.

13 Robert D. Bullard, *Dumping in Dixie: Race, Class, and Environmental Quality*, Boulder, CO: Westview Press, 1990, 10.

14 See, e.g., Nora Räthzel and David Uzzell (eds), *Trade Unions in the Green Economy: Working for the Environment*, Abingdon-on-Thames: Routledge, 2013.

the struggles against land dispossession in both Europe and the colonies, it can also be found in workplace resistance against the noxious impacts of the industrial revolution (for example, the smoke regurgitated by the coal-powered steam engine, or matchgirl exposure to white phosphorus).[15] At times, struggles against noxiousness within the workplaces (internal noxiousness) were broadened to target the noxiousness impacting upon the surrounding territories (external noxiousness), contributing to environmentalism turning into a form of mass politics.[16] From struggles for accident prevention to disputes over silicosis, asbestosis, lead poisoning or occupational cancers, there is a rich history of workplace environmentalism stemming from occupational safety and health demands.[17]

In this respect, Katrin MacPhee has usefully noted that a problem with 'neatly separating environmental activism and occupational health and safety struggles is the seeming surrender in doing so to bourgeois conceptions of environmentalism'.[18] That is, conceptions which predominantly concentrate on middle-class, 'post-materialist' and lifestyle politics. In opposition to these, environmental labour studies considers working-class, self-interested and production-oriented politics a more promising current of environmentalism. A common distinction is made here between 'moderate' and 'radical' strands of labour environmentalism.[19]

15 See Patrick Bresnihan and Naomi Millner, *All We Want Is the Earth: Land, Labour and Movements Beyond Environmentalism*, Bristol: Bristol University Press, 2023; Barbara Harrison, 'The Politics of Occupational Ill-Health in Late Nineteenth Century Britain: The Case of the Match Making Industry', *Sociology of Health and Illness* 17(1), 1995, 20–41.

16 Chad Montrie, *The Myth of Silent Spring: Rethinking the Origins of American Environmentalism*, Berkeley: University of California Press, 2018.

17 See, e.g., Elena Davigo, 'Il movimento italiano per la tutela della salute negli ambienti di lavoro (1961–1978)', PhD thesis, Università degli Studi di Firenze, 2018; Arthur McIvor, 'Guardians of Workers' Bodies? Trade Unions and the History of Occupational Health and Safety', *Labour History* 119, 2020, 1–30; Christopher C. Sellers and Joseph Melling (eds), *Dangerous Trade: Histories of Industrial Hazard across a Globalizing World*, Philadelphia: Temple University Press, 2012.

18 Katrin MacPhee, 'Canadian Working-Class Environmentalism, 1965–1985', *Labour / Le Travail* 74, 2014, 123–49, 130.

19 Dimitris Stevis and Romain Felli, 'Global Labour Unions and Just Transition to a Green Economy', *International Environmental Agreement* 15, 2015, 29–43.

The paradigm for moderate proposals is ecological modernisation: from this perspective, green growth spurred on by different amalgams of labour mobilisation and technological progress can make capitalism, workers' welfare and environmental preservation mutually compatible. Radical proposals, in contrast, call capitalist growth and capitalist technology – in short, capitalism – into question, holding that 'just transitions' will only be sustainable if they form part of a broader transition towards a classless society.

However, while the workplace is a strategic site of workers' leverage, 'job blackmail' acts as a constant force for circumscribing the ambitions of workplace-centred environmentalism. Employers can always threaten unions demanding cleaner production with the decommissioning of productive units and job losses, and just transition plans are equivalently often perceived as a risky endeavour. In fact, the workers employed in a polluting workplace often have good reasons to suspect that the newly created 'green' jobs might turn out to be of poorer quality relative to the old, 'grey' ones. Workplace-centred perseverance in struggles against external noxiousness is a sign of solid ethics, but it is not directly grounded in material interests and, therefore, can only be expected to go so far. Theorists who respond to a one-sided focus on the community with an equally one-sided focus on the workplace, such as Matthew Huber, ignore the fact that the realm in which workers experience most directly a material interest in eradicating external noxiousness is the community, not the workplace.[20]

As it is understood here, working-class environmentalism differs from environmental justice and environmental labour studies (in its early version) because of the central role it assigns to the problem of convergence between workplace and community organising.[21] As such,

20 Matthew T. Huber, *Climate Change as Class War: Building Socialism on a Warming Planet*, London: Verso, 2022.

21 While initially focused on trade unions in the industrial sector, environmental labour studies later expanded to include all types of workers and workers' organisations in their analysis, incorporating the insights of working-class environmentalism. See Nora Räthzel, Dimitris Stevis, and David Uzzell (eds), *The Palgrave Handbook of Environmental Labour Studies*, London: Palgrave Macmillan, 2021.

working-class environmentalism is not in opposition to either environmental justice or environmental labour studies; rather, it builds on both. As we will see in Chapters 2 and 4, drawing on ecofeminist theory, working-class environmentalism expands the understanding of who the workers are and how they organise.[22] This encompasses waged and unwaged workers, workplace-centred and community-centred mobilisations, and trade unions as well as other organisational forms, such as social movement organisations, political parties, community associations and informal networks.

In highlighting the necessity of workers to environmentalism and vice versa, both environmental justice and environmental labour studies represent significant progress when compared to earlier understandings of environmentalism as a merely middle-class 'post-materialist' value.[23] In fact, while workers are often depicted as environmentally regressive, they bear little responsibility for environmental degradation while also being highly affected by it. They therefore have a distinct *material* interest in pushing back against the ecological crisis, with their organisations becoming crucial collective actors in this effort. Indeed, different experiences of working-class organising have thrown spanners in the vicious cycle of noxiousness accumulation. And yet, on its own, workplace-centred environmentalism is either limited to internal noxiousness or idealistic (as it is not grounded in immediate material interests). On the other hand, community-centred environmentalism, on its own, lacks the structural power to change production from within. The limitations of both workplace-centred and community-centred working-class environmentalism when approached in isolation demand we consider the problem of convergence.

In dealing with this issue, my theoretical point of departure lies in *operaismo* (also known as workerism), a New Left current that emerged around the struggles of Italy's factory workers in the 1960s and was mainly disseminated through the journals *Quaderni rossi* (1961–66) and *Classe operaia* (1964–67). *Operaismo* was originally animated by

22 Barca, *Workers of the Earth.*

23 Ronald Inglehart, *The Silent Revolution: Changing Values and Political Styles Among Western Publics*, Princeton, NJ: Princeton University Press, 1977.

intellectuals such as Raniero Panzieri, Romano Alquati, Mario Tronti, Toni Negri and Sergio Bologna, and represented a rupture with the leadership of the Partito Socialista Italiano (Italian Socialist Party, PSI) and the Partito Comunista Italiano (Italian Communist Party, PCI). Markers of *operaismo* include an emphasis on the role of class struggle in shaping capitalist development and a focus on workers' autonomy in relation to party, union and state structures.[24] *Operaismo* owed much of its initial success to the fact that its theses seemed to be vindicated by the wave of working-class militancy that swept Italy in 1968–69 and took the official labour movement largely by surprise.[25]

This does not mean, however, that I follow the subsequent and well-known strand of '*post-operaismo*' as popularised by Michael Hardt and Toni Negri, although I respect and am indebted to this tradition.[26] A good way to understand the break between 'classic' *operaismo* and Negrian *post-operaismo* is to compare Raniero Panzieri's 'Surplus Value and Planning', published by *Quaderni rossi* in 1964, and Toni Negri's 'Crisis of the Planner-State', presented at the 1971 congress of the radical left organisation Potere Operaio (Workers' Power).[27] Panzieri and Negri died almost sixty years apart, the former in 1964 and the latter in 2023. Their fleeting collaboration in the early 1960s can be seen as a momentous watershed.

Operaista trailblazer Raniero Panzieri based his understanding of capitalist work and technology on his reading of Marx's *Capital*, stressing the notion of cooperation Marx offered there:

24 Gigi Roggero, *Italian Operaismo: Genealogy, History, Method*, Cambridge, MA: MIT Press, 2023 [2019].

25 Nanni Balestrini and Primo Moroni (eds), *The Golden Horde: Revolutionary Italy, 1960–1977*, Chicago: University of Chicago Press, 2021 [1988]; Steve Wright, *Storming Heaven: Class Composition and Struggle in Italian Autonomist Marxism*, London: Pluto Press, 2002.

26 See Michael Hardt and Antonio Negri, *Empire*, Cambridge, MA: Harvard University Press, 2000.

27 Antonio Negri, 'Crisis of the Planner-State: Communism and Revolutionary Organization' [1971], in *Books for Burning: Between Civil War and Democracy in 1970s Italy*, ed. Timothy S. Murphy, London: Verso, 2005, 1–50; Raniero Panzieri, 'Surplus Value and Planning: Notes on the Reading of *Capital*' [1964], in *The Labour Process and Class Strategies* (CSE Pamphlet no. 1), London: Stage 1, 1976, 4–25.

> Being independent of each other, the workers are isolated. They enter into relations with the capitalist, but not with each other. Their co-operation only begins with the labour process, but by then they have ceased to belong to themselves. On entering the labour process they are incorporated into capital. As co-operators, as members of a working organism, they merely form a particular mode of existence of capital.[28]

As Panzieri commented, workers' production in capitalism is shaped by capital: 'Cooperation in its capitalist form is the first and basic expression of the law of (surplus) value.'[29] On these bases, Panzieri developed a critique of the productivist version of Marxism upheld by the PCI mainstream in the post–Second World War years. In opposition to that, Panzieri did not cast the content of capitalist production as a neutral and rational technological base imprisoned by capitalist social relations, but, rather, as determined to a significant extent by capitalist social relations themselves: 'The relations of production are *within* the productive forces, and these have been "moulded" by capital.'[30]

Negri's approach, in contrast, was premised on his interpretation of Marx's *Grundrisse*, which pre-dated the first volume of *Capital* by roughly a decade. In the now oft-quoted 'Fragment on Machines', Marx imagined a stage of capitalist development in which:

> On the one side, it [capital] calls to life all the powers of science and of nature, as of social combination and of social intercourse, in order to make the creation of wealth independent (relatively) of the labour time employed on it. On the other side, it wants to use labour time as the measuring rod for the giant social forces thereby created, and to confine them within the limits required to maintain the already created value as value. Forces of production and social relations – two

28 Marx, *Capital*, vol. 1, 451.
29 Panzieri, 'Surplus Value and Planning', 12.
30 Ibid.

> different sides of the development of the social individual – appear to capital as mere means, and are merely means for it to produce on its limited foundation. In fact, however, they are the material conditions to blow this foundation sky-high.[31]

Here, we are much closer to the traditional Marxist version of historical materialism, in which the development of the technological base acts as an independent variable on the social superstructure. In the 'Fragment on Machines', in fact, Marx traced a tendency for the law of value – that is, the competition-enforced pressure for prices to be grounded in the average labour time needed to produce commodities, a pressure moulding capitalist work qualitatively – to fall into obsolescence due to scientific and technological progress.

Panzieri, however, correctly saw that the 'Fragment on Machines' is contradicted by Marx's later writings:

> In the quoted fragment, there is a model for a 'shift' from capitalism *directly* to communism. Against this, there are numerous passages in *Capital* and *Critique of the Gotha Program*. This problem will be the object of a detailed analysis in one of the future issues of *Quaderni rossi*.

This is footnote 76 in the original version of 'Surplus Value and Planning', but it was cut in the English translation referenced here. In any case, Panzieri died by cerebral embolism in the same year, aged forty-three, and *Quaderni rossi* would soon cease to be published.

Negri retained an *operaista* emphasis on capitalist technological development as a reaction to class struggles (an emphasis that, again, can also be found in *Capital*, volume 1), but nonetheless embraced the 'Fragment on Machines', underscoring that 'the development of the forces of production faces a barrier in the capitalist appropriation of

31 Karl Marx, *Grundrisse: Foundations of the Critique of Political Economy*, London: Penguin, 1973 [1939], 706.

wealth.'[32] Moreover, Negri saw the tendency of the law of value to fall into obsolescence as having become an actual 'presence':

> The crisis of 1929, or perhaps it would be better to say the moment at which, in response to the revolutionary socialist challenge and the October Revolution, the tendency passed to mass production as a means of destroying the conditions for worker organization . . . from this point, production becomes based on a general labor. The social character of production makes the product, from the outset, a general, social product. But in the passage we have reached today, the mystifications of this recomposition of capital and the state after 1929 are exhausted.[33]

According to this periodisation, in the Fordist phase of capitalist development, the workers were no longer isolated producers reunited by capital through the mediation of money, but instead already immediately social producers, which resulted in a 'disconnection between work and the general law of value.'[34] For Negri, communism was already creeping into the economic base, while capitalism lingered on in the political superstructure, attempting to rescue the law of value through 'desperate' and 'arbitrary' violence.

Negri was never a proponent of the neutrality of technological development, far from it. Nonetheless, a distinctive technological optimism is discernible in his approach. In Negri's schema, technology develops as a reaction to class struggle, but it does so in a way that eventually ends up undermining the discipline of the law of value on workers' cooperation in the labour process. If what emerges in Panzieri is a refusal of capitalist work *and* capitalist technology insofar as they are shaped by capitalist social relations, then in Negri we find a refusal of work *through* technology. The technological optimism that surfaces in 'Crisis of the Planner-State' can be recognised in Negri's later writings, as seen in his endorsement of left accelerationism.[35]

32 Negri, 'Crisis of the Planner-State', 21.

33 Ibid., 17.

34 Ibid., 23.

35 Antonio Negri, 'Reflections on the Manifesto for an Accelerationist Politics', *EuroNomade*, 6 March 2014, euronomade.info.

The thesis of the obsolescence of the law of value reappears in Hardt and Negri's texts on 'immaterial' labour.[36] Maurizio Lazzarato defined 'immaterial' labour as 'the labor that produces the informational and cultural content of the commodity'.[37] 'Immaterial' labour can thus be more easily understood as designating the intellectual component of work, which is also material as it emerges from a biological basis. Hardt and Negri theorised that the hegemony of 'immaterial' labour engenders networks of productive cooperation that tend to escape capital's control, resulting in 'increasingly autonomous forms of social production' that can provide the infrastructure for the future society.[38]

However, Hardt and Negri's claims about the supposed autonomy of 'immaterial' labour are problematic, as the law of value is enforced via market competition, regardless of whether this happens through direct capitalist supervision or through workers' internalised self-discipline. If we accept, with Panzieri, that the capitalist labour process is shaped by the law of value, then working-class autonomy as the political expression of workers' interests is not constructed through capitalist cooperation but against it, against capitalist work. In other words, autonomy is not situated in capitalist production, as the thesis of 'immaterial' labour suggests. Rather, when autonomy does emerge, it lies in workers' antagonism towards this very form of production.

Furthermore, the segmentation of the global working class means that reducing the contemporary composition of labour power to 'social and cognitive labor' does not do justice to its diversity.[39] While all work has always had a social (i.e. cooperative) and cognitive (i.e. intellectual) component, most members of the global working class are not employed

36 See, e.g., Michael Hardt and Antonio Negri, *Multitude: War and Democracy in the Age of Empire*, London: Penguin, 2004, 145.

37 Maurizio Lazzarato, 'Immaterial Labor', in Michael Hardt and Paolo Virno (eds), *Radical Thought in Italy: A Potential Politics*, Minneapolis: Minnesota University Press, 1996, 132–48, 132.

38 Michael Hardt and Antonio Negri, *Assembly*, Oxford: Oxford University Press, 2017, 209.

39 Hardt and Negri, *Assembly*, 193.

in jobs in which the intellectual component is dominant. The problem is not merely quantitative but also qualitative, as many important struggles – particularly in the Global South – have not been led by the 'cognitariat' but by workers at the margins of capitalist development. The thesis of the hegemony of 'immaterial' labour is thus also vulnerable to accusations of Eurocentrism.[40]

Post-operaismo, as a whole, has undoubtedly provided valuable contributions, updating class composition analysis to include the relative deindustrialisation of employment and diffusion of intellectual work in the Global North.[41] This book, however, proposes taking one step back and two steps forward in respect to Hardt and Negri's framework. The step back: a return to the classic *operaista* critique of capitalist work and technology. The first step forward: an articulation of *operaismo* with strands of feminist Marxism and Marxism from the Global South. The second step forward: an application of this theoretical wealth to the ecological question.

Of course, *operaismo* is itself full of ambiguities, internal diversity and blind spots. Its own renegade feminist strand – with authors such as Mariarosa Dalla Costa, Alisa Del Re, Silvia Federici and Leopoldina Fortunati – correctly and forcefully tackled such blind spots early on, subversively appropriating Tronti's concept of the 'social factory' to analyse reproductive work and the community as a site of class struggle. *Operaismo* is, therefore, far from sufficient on its own. When dealing with issues related to gender, race and the environment, in particular, it has to be complemented with other traditions. These necessarily come from people and places variously distant from the Italian factories, communities and universities where *operaismo* originated. To me, having grown up with it, *operaismo* simply provides a situated starting point from which to reach out to other perspectives.

40 See, e.g., Ramón Grosfoguel, 'Del imperialismo de Lenin al *Imperio* de Hardt y Negri: "Fases superiores" del eurocentrismo', *Universitas humanística* 65, 2008, 15–26.

41 See, e.g., Andrea Fumagalli, Alfonso Giuliani, Stefano Lucarelli and Carlo Vercellone, *Cognitive Capitalism, Welfare and Labour: The Commonfare Hypothesis*, Abingdon-on-Thames: Routledge, 2019.

In my view, one of *operaismo*'s most enduring legacies is its method of class composition analysis, according to which political interventions and organisational forms should constantly be updated to keep up with the objective and subjective transformations of working-class composition in different places and times.[42] Accordingly, *operaismo* is not inherently anti-union or anti-party, as the superficial clichés would have it, but offers a flexible approach that has inspired many different proposals and political interventions in the course of its history. The method is one of conducting workers' inquiries into the class composition existing in any given context, formulating political and organisational proposals, and testing their effectiveness through trial and error.[43] There is, therefore, a process of continuous exchange between the production of knowledge and political intervention.

The phrases 'technical composition' and 'political composition' were first coined by Romano Alquati in relation to the nature of the working class.[44] The *technical composition* is the 'objective' side, designating the ways in which workers are deployed, segmented and stratified as labour power in the workplace through different economic sectors, labour processes, commodity chains, etc. The *political composition* of the working class, on the other hand, indicates the extent to which workers as a class do *or do not* overcome their divisions to assert their common interests against capital. This is the 'subjective' side, made up of workers' forms of consciousness, struggle and organisation. Following the renegade feminist strand of *operaismo* inaugurated by Mariarosa Dalla Costa – which was in and against, in continuity and rupture with, earlier *operaismo* – the scholar-activists Seth Wheeler and Jessica Thorne usefully proposed an update to this framework by adding the *social composition*

42 Sergio Bologna, 'Class Composition and the Theory of the Party at the Origin of the Workers-Council Movement', *Telos* 13, 1972, 4–27.

43 Davide Gallo Lassere and Frédéric Monferrand, 'Inquiry: Between Critique and Politics', *South Atlantic Quarterly* 118(2), 2019, 444–56; Asad Haider and Salar Mohandesi, 'Workers' Inquiry: A Genealogy', *Viewpoint Magazine*, 27 September 2013, viewpointmag.com.

44 Romano Alquati, 'Composizione della classe: Una ricerca sulla struttura interna della classe operaia italiana', *Classe operaia* 13(1), 1965, 8–14.

of the working class.[45] This refers to the ways in which workers are reproduced in the community, for example through family, housing, welfare and health regimes.[46] The objective side of class composition is then bifurcated between technical composition (related to the workplace) and social composition (related to the community).

From this perspective, it is possible to analyse how the working class is also segmented in relation to environmental degradation.[47] Importantly for our purposes, *operaismo* stressed the duality of the working class both as a cog in the capitalist machine and as a spanner in its works.[48] The method of class composition analysis developed from these premises is then useful for making sense of the wide variety of stances historically taken by workers towards noxious production. These cover the full spectrum from Minamata's petrochemical labour union acting as a militia to attack fenceline community groups protesting mercury poisoning, to the Porto Marghera *operaista* group demanding an end to the 'production of death' from within the petrochemical complex that employed them.

Capitalist noxiousness

The term 'noxiousness', as used here, is a translation of the Italian word *nocività*, which indicates the property of causing harm. Through its deployment by the Italian labour movement, it came to refer to *production-induced harm against both human and non-human life*. Noxiousness, thus understood, goes beyond toxicity and accidents to include the hazards to mental and physical health engendered by the organisation of

45 See Anna Curcio, 'Marxist Feminism of Rupture', *Viewpoint Magazine*, 14 January 2020, viewpointmag.com.

46 Seth Wheeler and Jessica Thorne, 'The Workers' Inquiry and Social Composition', *Notes from Below*, 29 January 2018, notesfrombelow.org.

47 See Nick Dyer-Witheford, 'Struggles in the Planet Factory: Class Composition and Global Warming', in Jan Jagodzinski (ed.), *Interrogating the Anthropocene: Ecology, Aesthetics, Pedagogy, and the Future in Question*, Berlin: Springer, 2018, 75–103.

48 Tronti, *Workers and Capital*.

production. Such a concept of noxiousness also avoids a hard dualism between (human) health and the (non-human) environment, pointing to the interdependencies between the two.

That said, a distinction between the human and non-human components of ecology is still needed, particularly because the human species has, despite all the intra-human hierarchies, received better treatment from capitalist noxiousness than most non-human species. This is not to claim that humanity is an undifferentiated whole. Indeed, the political economy of health has provided abundant evidence of health inequalities structured along multiple stratification lines.[49] Similarly, as just discussed, environmental justice studies, political ecology and ecofeminism have shown how low-income, racialised and gendered subjects are disproportionately exposed to environmental degradation. Nonetheless, in many ways, noxiousness to humans was mitigated under capitalism through the dialectic between technoscientific development and struggles for health (as indicated by the global rise in average life expectancy).

And yet, the unprecedented scale of today's environmental devastation threatens the conditions of both human and non-human life on the planet. In recent decades, standards of capitalist production have improved, on the whole, thanks to health and environmental struggles. However, green techno-fixes have been offset by output growth and the *cumulative* nature of environmental degradation. It is the accumulation of noxiousness that has led to today's planetary ecological crisis, with its well-known predicaments including global heating, biodiversity loss, soil depletion and persistent organic pollution. Greenhouse gases accumulate in the atmosphere, dioxins pile up in food chains, desertification and deforestation march on, and plastics are amassed in the oceans. Faced with all this, moderate technological improvements remain wholly inadequate. Indeed, this section argues that the '*how much*', the '*what*'

49 See, e.g., Jaime Breilh, 'Critical Epidemiology in Latin America: Roots, Philosophical and Methodological Rupture', in Jordi Vallverdú, Angel Puyol and Anna Estany (eds), *Philosophical and Methodological Debates in Public Health*, Berlin: Springer, 2019, 21–46; Lesley Doyal, *The Political Economy of Health*, London: Pluto Press, 1979.

and the '*how*' of capitalist production are inherently geared towards generating and aggravating the ecological crisis.

Noxiousness in general is not an exclusive feature of capitalism. Feudalism and slavery, for example, also gave rise to substantial levels of production-induced fatalities, injuries, diseases, pollution and extinctions. Nonetheless, the environmental degradation caused by previous modes of production pales in comparison to the ecological crisis generated by capitalism. What is the specific difference inherent to the capitalist form of noxiousness that has made an ecological crisis of this scale possible?

By definition, production always has a role in generating noxiousness, but many distinctions can be made. Noxiousness can be, and usually is, released at all stages of a product's life cycle, from the extraction of its raw materials to its intermediate processing, distribution, final consumption and waste disposal. Indeed, since fossil fuels are the main driver of global heating, significant noxiousness is released in any process of production or consumption powered by hydrocarbons. When production and consumption happen at the same time, as is the case for most services, noxiousness is released simultaneously at the points of production and consumption. Noxiousness at the point of production can, in turn, be internal or external. 'Internal' here refers to damage within the work environment (for example, employees' exposure to toxic gases), while 'external' denotes impacts on the outside (for example, a factory polluting the air, water and soil around its premises). While internal noxiousness is, by definition, localised, external noxiousness can be local, global or both. Many toxic substances only impact upon lives positioned within a limited range from the emission point, but greenhouse gases contribute to global heating no matter where their source is located. Another distinction is that between tangible noxiousness (emissions, physical hazards, noise, etc.) and organisational noxiousness constituted by the harmful elements of certain work regimes (night shifts, fast-paced work, monotony, isolation, etc.).[50]

50 See Ivar Oddone (ed.), *Ambiente di lavoro: La fabbrica nel territorio*, Rome: Editrice Sindacale Italiana, 1978.

All these distinctions – noxiousness at the points of production and consumption, internal and external, local and global, tangible and organisational – are useful. However, they do not directly help to answer our question about the specificity of the capitalist form of noxiousness. In fact, all these types of noxiousness can also occur in non-capitalist modes of wealth creation, but only *capitalist* noxiousness has generated the ecological crisis. Even the distinction between tangible and organisational noxiousness does not map onto the capitalist versus non-capitalist distinction. This is because, on the one hand, harm linked to the organisation of work can occur in non-capitalist modes of production too, and, on the other hand, tangible noxiousness under capitalism is (just like organisational noxiousness) shaped by capitalist social relations. What then is the specificity of capitalist noxiousness?

To put it biblically, in the beginning was the class struggle, understood here in an intersectional and non-reductionist way, accounting for its racialised, gendered and further dimensions. That is, the fully fledged domination of capital begins with class struggle in the form of the separation of direct producers from the means of production.[51] The workers were pulled apart from the land – their 'natural workshop', as Marx put it in the *Grundrisse* – in the first place, and thus from nature itself.[52] Marx called this process 'primitive accumulation' and noted that it occurred historically through both domestic enclosures (the privatisation of common lands and the expulsion of the peasants from them) and imperial conquest (the subjugation of indigenous peoples and the appropriation of their territories).[53] Initially, 'free' wage labour was not the dominant regime of work control – as it was disproportionately confined to white, male workers – and, even today, it is far from being the only one, as demonstrated by the persistence of modern slavery, unwaged work,

51 Werner Bonefeld, *Critical Theory and the Critique of Political Economy: On Subversion and Negative Reason*, London: Bloomsbury, 2014; Silvia Federici, *Caliban and the Witch: Women, the Body and Primitive Accumulation*, New York: Autonomedia, 2004.

52 Marx, *Grundrisse*, 471.

53 Marx, *Capital*, vol. 1, 915–16.

bogus self-employment, etc. Nevertheless, primitive accumulation ensured the dependency of a growing share of the global workforce on production for the market, as increasing numbers of people could no longer rely on self-consumption.

The imperial destruction of pre-colonial modes of production, and the ensuing subordination of colonised labour to commodified production, was not a fact of 'civilised' commercial expansion but one of extremely brutal class struggle.[54] As Héctor Alimonda writes, the revelation of coloniality as the necessary and constitutive darker side of capitalist modernity is

> a point of rupture equivalent to that of Chapter 24 of *Capital* [vol. 1], when Marx – attacking the sugar-coated fairy tales of political economy on the natural origin of the categories of mercantile economy – introduces into his narrative, just like an eruption, the historical analysis of primitive accumulation, through which capital is born and constituted by violence, 'in letters of blood and fire'.[55]

This fundamental separation between workers and the means of production must constantly be recreated, along with the pulverisation of production across a multiplicity of firms competing among themselves, if capitalism is to persist. The ensuing prevalence of the commodity form of wealth severs the workers from non-human nature and divides capitalist work into direct commodity production and reproduction. Within this schema, reproductive work, entailing the indirectly commodified creation and maintenance of an employable workforce, is subordinated to commodity production.[56] By contrast, in pre-capitalist societies wealth creation was mostly unified under small-scale self-consumption, while in a hypothetical socialist society it would be unified under multiple layers of democratic planning.

54 See José C. Mariátegui, *José Carlos Mariátegui: An Anthology*, New York: Monthly Review Press, 2011.

55 Alimonda, *La naturaleza colonizada*, 26.

56 Mariarosa Dalla Costa and Selma James, *The Power of Women and the Subversion of the Community*, Bristol: Falling Wall Press, 1972.

In capitalism, non-human nature enters the labour process as means of production (inputs and machines) from which the workers are separated. Zooplankton and algae fossils, for example, naturally become oil and gas and are then turned by human work into fuels and raw materials for the petrochemical industry. They thus mutate into plastics, such as the toxic 'forever chemicals' per- and polyfluoroalkyl substances (PFAS) used to make, among other things, microchips.[57] Workers' relationship with nature occurs through an alienated labour process brought into existence by capitalist investment. Even in unwaged reproductive work, whose meagre means of production are usually owned by the workers, work and its relationship to nature are conditioned by the necessity of reproducing an employable work-force for capital. Anti-capitalist struggle is, then, a struggle against separation, a struggle for the convergence, on new bases, of what has been separated.

Dispossession and commodification likewise generate the contradiction between wealth (or use value) and value. Wealth is created under all modes of production, whether it be under the coercion of a feudal overlord who appropriates a share of the product, or, in principle, through democratic planning about what, how and how much to produce. In non-capitalist societies, work mainly produces wealth that is not exchanged on the market. Conversely, under capitalism, work mostly produces wealth in the form of marketable commodities. The price of such commodities is a reflection of their value. That is, the amount of average labour time (or abstract labour) needed to make them.[58] This is the law of value: constantly exerting the competitive pressure of abstract labour (the social average constituted in production but enforced through market exchange) on concrete labour.[59]

57 Cheng Ting-Fang, 'The Crackdown on Risky Chemicals that Could Derail the Chip Industry', *Financial Times*, 21 May 2023.

58 Marx, *Capital*, vol. 1, 138–62.

59 See Søren Mau, *Mute Compulsion: A Marxist Theory of the Economic Power of Capital*, London: Verso, 2023.

Value is seen here as being both conditional on and constituted by class struggle.[60] Class struggle is what makes and keeps workers separate from the means of production, and it is therefore the foundation of the law of value.[61] Moreover, as a social average, abstract labour is constituted by struggles over the intensity of work, which also act as an incentive to technological development. These conflicts might be expressed through the eventful macro-struggles of organised mass mobilisation by politically conscious workers, but they are also present in the everyday, capillary and politically unconscious micro-struggles to improve one's life within the labour process.[62] Socially necessary labour time, the average labour time that constitutes value at any given moment, is the net result of such an antagonism. The law of value is thus conditional on struggle (on dispossession and its reproduction in time), as well as constituted by struggle (through the conflictual determination of abstract labour as average labour time).

The commodification of labour power and nature subordinates the creation of wealth to impersonal market competition and the profit imperative of capital accumulation. In other words, the 'how much', the 'what' and the 'how' of production escape conscious control and are ruled by forces that act 'behind the backs of the producers'.[63]

Regarding the 'how much', capitalism entails an ever-growing expansion of material output. In fact, as the capitalist economy is driven by profit, money is invested only if a gain in value on the initial sum is expected. Value is not a measure of material quantities but of the average labour time necessary to produce the commodities made of such material quantities. While commodities are usually not sold at their value for manifold reasons, prices are ultimately rooted in value and are thus also

60 Harry Cleaver, *Rupturing the Dialectic: The Struggle Against Work, Money, and Financialization*, Oakland, CA: AK Press, 2017.

61 Werner Bonefeld, 'On Postone's Courageous but Unsuccessful Attempt to Banish the Class Antagonism from the Critique of Political Economy', *Historical Materialism* 12(3), 2004, 103–24.

62 John Holloway, *In, Against, and Beyond Capitalism: The San Francisco Lectures*, Oakland, CA: PM Press, 2016.

63 Marx, *Capital*, vol. 1, 135.

a temporal category, not a physical one. It is then necessary to distinguish between 'material output' and 'price output'. Material output refers to physical quantities, say 10 tonnes of iron, and it is thus only comparable for units of the same product. Price output refers to the market price of a product at any given point, say 100 dollars per tonne of iron. This measure makes all commodities comparable and it ultimately depends on the average labour time necessary for their production.

Market competition and class struggle generate a constant pressure to raise labour productivity – that is, to make more stuff in less time – in order to reap the surplus profits that come from producing faster than the social average.[64] If, by introducing a new robot, a smartphone assembly company manages to produce more devices than its competitors in the same amount of time, it will incur lesser costs per device while still being able to sell its smartphones at a price similar to that of other firms. It will thus pocket an extra profit. However, once a productivity-raising innovation becomes widely used across firms, these surplus profits vanish. When the robot becomes a standard feature of the smartphone industry, costs per device will be equalised, and the extra profits once enjoyed by the innovating firm will evaporate.

Cumulative productivity gains thus endlessly widen the distance between average labour time (that is, value) and material output (including waste). In other words, value per unit of material output constantly declines. In our example, the average labour time embodied in a smartphone becomes shorter and shorter. At the end of an innovation race, the total value produced by society increases only to the extent that more units of average labour time are drawn into the system, even if material output continues to grow by a larger proportion due to the expanded productivity of labour. The endless self-expansion of value necessary for jobs to be created and maintained therefore corresponds to a much faster growth in the material throughput of the global economy.

This leads to the 'Jevons Paradox', in which increased efficiency does not result in less consumption. Rather, as the value of given quantities of a product falls, making it more affordable, demand grows along with

64 Ibid., 429–38.

material output – and grow it must if profitability is to be protected. As shown by Kohei Saito's reading of Marx, there is an inherent contradiction between the value form of capital and the material content of nature: 'Capital contradicts the fundamental limitedness of natural forces and resources because of its drive toward infinite self-valorization.'[65] Furthermore, the distribution of wealth generated as a by-product of capital accumulation must be unequal enough to ensure the continuing subordination of the global working class to profitable production, and thus to capital's infinite need for growth.

However, the problem is not merely a quantitative one. To imagine so would risk concluding that the provision of dignified living standards to the whole of humanity is incompatible with the long-term reproduction of other species. The fact is that the profitability of a company results not only from efficiency, but also from the ability to make things people will buy. If an automotive company becomes able to produce sports utility vehicles (SUVs) faster than its competitors, its profits, other things being equal, will increase. However, if a sudden change in consciousness makes motorists less inclined to poison their fellow citizens' air just for the sake of having a massive car, demand for SUVs will collapse and the company in question will be making losses instead of profits. Yet, choosing what to consume on the market – the 'what' of production – is not the same as doing it via democratic planning. As Murray Bookchin noted, in capitalism:

> Production and consumption, in effect, acquire suprahuman qualities that are no longer related to technical development and the subject's rational control of the conditions of existence. They are governed instead by a ubiquitous market, by a universal competition not only between commodities but also between the creation of needs – a competition that removes commodities and needs from rational cognition and personal control.[66]

65 Kohei Saito, *Karl Marx's Ecosocialism: Capital, Nature, and the Unfinished Critique of Political Economy*, New York: Monthly Review Press, 2017, 171.

66 Murray Bookchin, *The Ecology of Freedom: The Emergence and Dissolution of Hierarchy*, Palo Alto, CA: Cheshire Books, 1982, 53.

Market choices are intrinsically individualist and short-term, while democratic planning is collective and potentially far-sighted. Through democratic economic planning, the majority of citizens may decide that SUVs do not make sense except for limited uses. In a world of atomised consumption, individually opting for a small car, or (when possible) no car at all, is an ethical choice to be widely encouraged. However, individual consumer choices are, unfortunately, close to irrelevant in the grand scheme of things, and people know it. The core of the problem, then, lies in collectively changing the social relations that mould both production and consumption.

Within the workplace, capitalist production, like the capitalist product, has two faces. Just as the commodity has both a use value and an exchange value, capitalist production includes both the labour process and the valorisation process.[67] The labour process refers to the production of wealth in general. The valorisation process is the production of surplus value (that is, the value produced by the workers beyond their own wage costs), and thus of profits, through the exploitation of the workforce. In capitalism, the concrete characteristics of the labour process – the 'how' of production, including its noxiousness – are shaped in alienated form by the valorisation process. The profit imperative is experienced by the worker as a subtraction of meaning from her activities, a constant need for speed and a pressure to economise on the instruments of work.

The technology deployed in the valorisation process is itself not a neutral instrument, because *capitalist* technology must secure the discipline of the workforce.[68] The most typical example is the assembly line, which, by fragmenting the labour process into simple, repetitive tasks and regulating their pace, forces upon the workers a quantitative overload (speed) and a qualitative underload (monotony) which is good for profitability but detrimental to their well-being.[69] Out of the full range of

67 Marx, *Capital*, vol. 1, 283–306.

68 David F. Noble, *America by Design: Science, Technology and the Rise of Corporate Capitalism*, New York: Knopf, 1977.

69 See Asa C. Laurell and Mariano Noriega, *La salud en la fábrica: Estudio sobre la industria siderúrgica en México*, Mexico City: Era, 1989.

possible developments, the technologies actually deployed in the capitalist labour process are the outcome of concrete struggles and are, in this sense, a reflection of the societal balance of power. Nonetheless, they must always be compatible with the profit imperative. Noxiousness is then shaped, and limited, by health and environmental struggles. But the dominance of the valorisation process over the labour process ensures that workers must still operate under different types of risk to both physical and mental health despite the preventive measures technically available.[70]

In the last instance, the law of value acts as an impersonal force shaping the 'how much', the 'what' and the 'how' of production. For profitability to be maintained, material output must be constantly expanded in the form of marketable commodities that meet individual consumer demand, commodities that are produced by disciplined workers in the shortest possible time. To paraphrase Marx, under capitalism the accumulation of wealth is at the same time the accumulation of relative misery and absolute noxiousness, hence its antagonistic character.[71] However, dispossession ties the workers' very livelihoods to their employer's profitability. If workers undermine the competitiveness of their companies – for example, by demanding the discontinuation of particularly noxious products without a lucrative alternative in sight – they can expect to lose their jobs. If workers cripple the bases of sustained economic growth – asking for state policies in defence of the environment that would weaken investor confidence – they can expect unemployment to rise.

Going back to our initial question, capitalist noxiousness is historically specific because it is shaped by the struggle-constituted law of value, while non-capitalist noxiousness merely occurs in the process of producing wealth. The law of value – with its profit imperative and its unconscious, and thus irrational, determination of production and distribution – makes production quantitatively unbound and qualitatively blind to

70 See Karl Marx, *Capital: A Critique of Political Economy*, vol. 3, London: Penguin, 1991 [1894], 180.

71 See Marx, *Capital*, vol. 1, 799.

environmental degradation. The latter is then constantly expanded in capital's 'metabolic rift', the widening fracture between capitalist production and its natural substratum.[72] On a planet where the best hydrocarbon reserves are depleted and the atmosphere is packed with CO_2, fossil fuel extraction can only happen at the cost of ever more destructive practices such as fracking.[73] Similarly, the extraction of profits in a society where it takes so little labour time to produce enormous amounts of wealth can only happen at the cost of ever more damaging environmental destruction. Hence the capitalist essence of the current ecological crisis.

Can it then be said that capitalism is intrinsically noxious? The answer is political and rests on a counterfactual proposition: is it possible to conceptualise an alternative mode of production generating much less noxiousness? A mode not driven by competition and profit, but by the cooperative and long-term satisfaction of socioecological needs through the creation of different goods with different technologies; a mode coupled with a more egalitarian distribution of wealth that can reconcile human health and environmental preservation. If this is at least possible in principle, then capitalism is indeed a noxious system.

From noxious deindustrialisation to working-class environmentalism

The first half of this book interprets our current predicament by relating the ecological crisis to Marx's 'general law of capitalist accumulation', according to which (other things being equal) capitalist development tends to simultaneously increase technological capabilities, on the one hand, and employment precarity, on the other.[74] This law thus has two

72 See Kohei Saito, *Marx in the Anthropocene: Towards the Idea of Degrowth Communism*, Cambridge: Cambridge University Press, 2023; Foster, *Marx's Ecology*.

73 Fracking, or hydraulic fracturing, is a non-conventional method of oil and gas extraction based on the injection of pressurised liquids into bedrock formations. In addition to its contribution to global heating, it is associated with increased risk of earthquakes and water contamination.

74 Marx, *Capital*, vol. 1, 762–870.

faces. The first concerns capital composition, the rising organic composition of capital, which is analysed in Chapter 1. The second concerns class composition, the rising precarity of the working class, which is the focus of Chapter 2. The second half of the book then looks for some of the social forces that might have an interest in changing this state of affairs consisting of increasing noxiousness and precarity. Chapter 3 argues that, due to the unevenness of the international division of labour and noxiousness, the working class in the Global South has a more direct interest in challenging the incumbent socioecological hierarchy. Furthermore, Chapter 4 proposes that community-centred interests can form the material bases for a working-class environmentalism that is geared towards recomposing workplaces and communities to face the ecological crisis.

Chapter 1 discusses what I call 'noxious (employment) deindustrialisation'. At the planetary level, this refers to a fall in the global industrial employment share coupled with the ecological crisis engendered by industrialised production, including the deployment of factory-produced machinery and inputs in non-industrial sectors and in final consumption. Locally, it means employment deindustrialisation, and a parallel rise in precarity, in areas where significantly noxious industries are still operating. This is illustrated through a case study of Grangemouth, Scotland's petrochemical capital.[75]

The chapter begins with a theoretical exploration of capital composition, discussing how the antagonistic co-evolution of class struggle and capitalist technology has resulted in the current configuration of advanced automation, dwindling industrial employment shares and accumulating noxiousness. It then goes on to provide statistical evidence for these claims on a global level. When noxious deindustrialisation takes place locally, fenceline communities remain exposed to significant levels of cumulative industrial noxiousness (despite improvements due to green techno-fixes), while reaping meagre benefits in terms of

75 Lorenzo Feltrin, Alice Mah and David Brown, 'Noxious Deindustrialization: Experiences of Precarity and Pollution in Scotland's Petrochemical Capital', *EPC: Government and Policy* 40(4), 2022, 950–69.

industrial jobs. Following the heyday of 'Boomtown Grangemouth' from the 1950s to the 1970s, for example, local employment in the petrochemical industry has gradually declined due to a combination of automation, rising qualification barriers and associated long-range recruiting, and outsourcing to a partially itinerant workforce. For the Grangemouth community, this trend has led to the current situation of employment deindustrialisation coupled with continuing exposure to the socioenvironmental damage and hazards engendered by operating petrochemical plants.

Acknowledging noxious deindustrialisation has implications for how we think about the so-called 'jobs versus environment dilemma'. That is, the notion that if something is to be gained in terms of good jobs, something must be lost on the side of health and the environment. Such a dilemma is particularly relevant to large-scale industry, which is seen both as an ideal-typical provider of secure, well-paid jobs and an ill-famed environmental felon. The understanding of the jobs versus environment dilemma as a zero-sum game has rightly been criticised as too pessimistic by advocates of 'just transitions', who refer to positive-sum games in which industry could be transformed to be sustainable and provide good jobs at the same time (although the social and political transformations needed for this to happen are often underestimated). From a different angle, however, the zero-sum game position is also too optimistic. This is precisely because it fails to take into consideration the reality of noxious deindustrialisation: a negative-sum game in which industrial job losses and environmental degradation progress simultaneously.

Chapter 2 explores the implications of the 'general law of capitalist accumulation' on the side of the global working-class composition, bringing in the example of precarious youths in Tunisia's Kerkennah archipelago and their social movement against the energy corporation Petrofac. Productivity gains in a context of slow economic growth mean that factory job losses are not completely recuperated in the service sector at comparable relative wages and conditions. This, coupled with the proletarianisation of the peasantry, reproduces and expands the ranks of what Marx called the surplus population, a phenomenon now more widely known as employment precarity. The chapter advocates for a broad notion of working-class composition, seeing the surplus population – since it is made up of workers dispossessed from the means of production – as a segment of the working class. To avoid any confusion, I term this segment the 'surplus working class'. Its precarity is defined by its dependence on, but irregular access to, a monetised income.

Precarity is much more widespread and severe in the Global South than in the Global North, and the former is also disproportionately exposed to the impacts of the ecological crisis. This is hardening what Desmond Tutu called a planetary 'climate apartheid', whereby the racialised segments of the global working class most vulnerable to the ecological crisis are restricted in their ability to move from sacrificed to safer zones. The apartheid imposed on the Palestinian people by the Israeli government, with its ongoing genocide, is currently the most acute and devastating manifestation of a much broader phenomenon. The prospect of masses of surplus workers excluded from core production and left to face the environmental destruction wreaked by that very production is a gloomy scenario, but it is already part of our world – as demonstrated by the high numbers of migrants attempting to reach Europe in inhumane conditions and at the risk of a tragic death.

In Tunisia, the decline of agricultural employment and stagnation of industrial employment engendered the expansion of a private service sector characterised by low wages and high precarity. In turn, this transformation of the Tunisian working-class composition ignited both a massive wave of external migration towards Europe and a cycle of internal struggles for secure employment and local development.[76] The employment fragmentation and insecurity of the surplus working class constitutes a significant obstacle to workplace organising, which is why the struggles of precarious youths are often community-centred: based on communal solidarities and organised in spaces of reproduction. In Kerkennah, these mobilisations targeted gas extraction in an archipelago

76 Lorenzo Feltrin, 'The Struggles of Precarious Youth in Tunisia: The Case of the Kerkennah Movement', *Review of African Political Economy* 45(155), 2018, 44–63.

that is already feeling the effects of global heating and its consequent rising sea levels and biodiversity loss.

Chapter 3 deals with the planetary unevenness of noxious deindustrialisation in a world stratified by colonial hierarchies, focusing on the 'sacrifice zone' of Ventanas on the coast of central Chile. Capital accumulation, while globally unified, is not globally homogeneous, as demonstrated by the persistent divergence between the Global South and Global North. Even if broadly similar trends can be identified in different parts of the world, each local experience of noxious deindustrialisation has its own specificities. A key element here is the mode of insertion of the affected area in the 'international division of labour and noxiousness', that is, the global socioecological hierarchy constituted by interconnected differentials in technological capabilities, wage levels and environmental degradation.

While global unevenness is a constant of capitalism, its concrete manifestations are in constant transformation. The 'old' international division of labour and noxiousness was predicated on the disproportional extraction of raw materials from the colonies for industrial processing in the metropolis. However, since the neoliberal turn of the 1970s, most of the world's factory employment was shifted to low-wage countries in the Global South, particularly in Asia. There are many Global South countries, nonetheless, particularly in Africa, South America and the Middle East, that are still characterised by extractivist economies, dependent on rent-bearing primary or 'quasi-primary' product exports. The current struggle over the international division of labour between the West and China has also been shaped by ecological transition attempts. The neoliberal, market-led version of the 'ecological transition from above' – that is, a vision for an ecological transition that keeps the profit imperative at the centre – underwent a long crisis after the 2007 financial crash. As a consequence, between 2020 and 2024, the United States and the European Union veered to a 'green' plan of capital, characterised by more direct state interventionism. However, this has also failed to stop China from narrowing its technological distance from the West. We are thus now seeing, in the latter, a shift to a 'white' plan of capital. State interventionism remains, but the ambitions at an ecological transition are dropped and, with the rise of the far right, a more explicit white supremacism is embraced.

In countries inserted in the international division of labour and noxiousness through extractivism, of which Chile is a prime example, the simple closure of large-scale polluting manufacturing industries deepens the dependence of their economies on primary product exports. However, the permanence of such industries often results in the perpetuation of 'sacrifice zones', where communities are exposed to severe noxiousness while receiving only meagre socioeconomic benefits in return. By analysing mobilisations over noxious deindustrialisation around the copper smelting and refining factories of Ventanas, this chapter shows how working-class environmentalism in the Global South can take the form of a struggle to break away from the incumbent division of labour and noxiousness.[77] This is particularly relevant in the context of today's energy expansion and scramble for 'critical minerals', and the related proliferation of armed conflicts.

Chapter 4 argues that community-centred interests, grounded in the defence of workers' conditions of reproduction, provide a material basis for a politics of working-class environmentalism. On a local level, as shown in the case of the industrial area of Porto Marghera in Northeastern Italy, noxious deindustrialisation coupled with long-distance commuting has disembedded the workforce of large-scale industry from the communities around the polluting factories. This increases the likelihood of community–industry conflicts and intra-working-class tensions between those employed in large-scale polluting industries (but living far from them) and those living nearby (but working in other sectors). The community is seen here as a site of class struggle through an expanded conceptualisation of the working class, work and working-class interests. This frame underpins an 'ecological turn' in class composition analysis, which proposes that considering the differential dependencies on and exposures to environmental degradation can help inform

77 Lorenzo Feltrin and Gabriela Julio Medel, 'Noxious Deindustrialisation and Extractivism: Quintero-Puchuncaví in the International Division of Labour and Noxiousness', *New Political Economy* 29(2), 2024, 173–91.

platforms of demands aiming to articulate workplace and community interests.

Porto Marghera was the site of an early experience of working-class environmentalism during Italy's Long 1968.[78] Here, the Porto Marghera *operaista* group waged an intervention in struggles against noxiousness that can be characterised by four elements: the thesis of the inherent noxiousness of capitalist work; an antagonist-transformative approach to capitalist technology; the connection between workplace and community struggles; and the demand for working-class control of 'what, how and how much to produce'. After the defeat of Long 1968, deindustrialisation kicked in while public awareness of industrial pollution rose, such that since the 1990s a divergence emerged between workplace-centred and community-centred struggles over chlorine-based petrochemical production in the area.[79] While the rival mobilisations limited damage to health and the environment on the one hand, and chlorine workers' livelihoods on the other, chlorine-based production in Porto Marghera was terminated without full environmental remediation and without the relocation of all its workers to comparable jobs. The chapter concludes that a convergence between workplace and community organising is a critical step in the construction of alternatives to the jobs versus environment dilemma.

The book's conclusion suggests three key elements that an 'ecological transition from below' – that is, an ecological transition in the interest of the working class, intersectionally understood – would have to strive for, namely: decommodification of production and nature, wealth redistribution, and a more balanced international division of labour. Pushing back against the frontier of commodification is a necessity for liberating increasing shares of wealth creation from the profit imperative and orienting this production towards collective planning for the long-term

78 Lorenzo Feltrin and Devi Sacchetto, 'The Work-Technology Nexus and Working-Class Environmentalism: Workerism versus Capitalist Noxiousness in Italy's Long 1968', *Theory and Society* 50(5), 2021, 815–35.

79 Lorenzo Feltrin, 'Situating Class in Workplace and Community Environmentalism: Working-Class Environmentalism and Deindustrialisation in Porto Marghera, Venice', *Sociological Review* 70(6), 2022, 1141–62.

satisfaction of human and non-human needs. Working less and better would be one way of achieving degrowth in the material throughput of society, at least in the Global North, while also improving quality of life. Yet wealth redistribution and a more balanced international division of labour are vital to make cleaner production socially sustainable and to reduce the global stratifications that, by keeping the working class divided, hold back the fight for decommodification. The following pages, then, explore the possibilities for, and obstacles to, the convergences between workplace and community struggles that are needed to move forward in such a direction.

1

Automation and Noxious Deindustrialisation: The Political Composition of Capital

In 1979, *Lavoro zero*, the journal of the Porto Marghera *operaista* group, published a special issue on science and technology. In an article titled 'The Two Sciences' they noted that 'when thinking about science, many people cannot help visualising it as the machinic apparatus dictatorially regulating their working day'.[1] This reflected the understanding that in capitalism there is a qualitative aspect to technological development, geared towards the control of the workforce. And yet, despite all the despotism embodied by capitalist industrial technology, a sort of 'smokestack nostalgia' has been observed among many deindustrialised communities.[2] Such nostalgia has played a major role in recent public debates in the Global North, including the reactionary Make America Great Again project to rewind history to the times when the US was the world's mightiest industrial powerhouse, with the heyday of post–Second World War industrial growth being widely remembered as a time of rising prosperity and security.[3]

1 *Lavoro zero*, 1979, 30.

2 Jefferson Cowie and Joseph Heathcott, *Beyond the Ruins: The Meanings of Deindustrialization*, Ithaca, NY: Cornell University Press, 2003.

3 Lawrence J. Broz, Jeffry Frieden and Stephen Weymouth, 'Populism in Place: The Economic Geography of the Globalization Backlash', *International Organization* 75, 2021, 464–94.

However, today's open factories no longer ensure such 'splendid fortunes and progressive pace', to echo Giacomo Leopardi's ironic expression.[4] In fact, successful industrial production in a given area is often associated with declining shares of factory jobs in total employment, while nearby communities remain exposed to industrial noxiousness. This paradox recalls the depictions of sci-fi dystopias in which out-of-control technologies turn against human and non-human life.[5] Technological imaginaries can, however, also point towards a different direction. For instance, the Porto Marghera *operaista* group illustrated their special issue with drawings by Moebius and other sci-fi artists, depicting inventive combinations of pre-capitalist and futuristic artefacts, indicating the possibility for a second, alternative science and technology.

I define noxious deindustrialisation as employment deindustrialisation (falling shares of industrial jobs in total employment) in areas where significantly noxious industries are still active. 'Noxious deindustrialisation' is thus, more precisely, 'noxious employment deindustrialisation', but I use the shortened phrase for the sake of brevity. In the statistical categories used by the International Labour Organization (ILO), 'industry' includes the utilities and construction sectors, while factory-based production falls under the heading of 'manufacturing'. However, construction is hardly internationally tradable, it is less amenable to automation compared to large-scale industry, and it tends to offer more precarious employment. These characteristics differentiate it from manufacturing and bring it closer to many services, which is why a narrower definition of industry is preferred here. In Marx's writings, by contrast, the terms 'industry' and 'industrial' are used both in a broad sense, referring to any productive sector, and in a strict sense, referring to factory-based production. Unless otherwise stated, when discussing industry I use the term in its strict sense, as factory-based production.

4 Giacomo Leopardi, 'The Genista or the Flower of the Desert' [1845], in *Giacomo Leopardi: Collected Works*, Bexhill-on-Sea: Delphi Classics, 2019.

5 Andrew Nette and Iain McIntyre (eds), *Dangerous Visions and New Worlds: Radical Science Fiction, 1950 to 1985*, Oakland, CA: PM Press, 2021.

The main driver of noxious deindustrialisation is industrial automation in a context of slow output growth. Obviously, 'automation' does not refer to the disappearance of the workforce from the labour process, something that has never happened. I rather use 'automation' in its broad sense, simply as a synonym for technology-induced productivity gains. The more industrial productivity increases faster than industrial output, the more workers are expelled from industrial production. Insofar as automation is needed for keeping factories competitive, employment deindustrialisation is often a necessary condition for maintaining industries in a given place. In fact, those factories that fall behind in the productivity race – that is, beyond the point at which they can compensate for their technological disadvantage by lowering wages and environmental standards – are doomed to bankruptcy and closure. Noxious deindustrialisation thus does not refer to closing factories, but to operating factories that require less living labour than they used to, relative to material output and total employment. It therefore differs from both the toxic legacies of prior production (such as long-term contamination of soils and waters) and the social disruption (such as underemployment and mental health disorders) caused by changing employment patterns *after* factories have closed.[6] Under noxious deindustrialisation, labour gets shed, but significantly noxious factories remain open and thus pollution continues.

I first began to elaborate the concept of noxious deindustrialisation during my fieldwork for the Toxic Expertise project in Grangemouth, a Scottish town whose landscape is dominated by the INEOS refinery and petrochemical complex. Just like any other visitor to the town, I was impressed by the sheer magnitude of the industrial infrastructure, recalling Alessandro Portelli's notion of an industrial sublime: 'A city of fire, of epic size, in which huge machines move fantastic objects between iron and fire, and where fear, beauty and wonder weave into a modern form

6 See Steven High, Lachlan MacKinnon and Andrew Perchard (eds), *The Deindustrialized World: Confronting Ruination in Postindustrial Places*, Vancouver: UBC Press, 2017; Alice Mah, *Industrial Ruination, Community and Place: Landscapes and Legacies of Urban Decline*, Toronto: University of Toronto Press, 2012.

of what Romantic poets and philosophers, from Kant to Blake, called "the sublime".[7]

Yet, once I started speaking with the locals, I was equally struck by their stories of how just about everything in the town seemed to have 'gone downhill' since the late 1970s. Participants reported a decline in industrial jobs, prosperity, population, community cohesion, mental health and public services, which mirrored accounts of factory closures I had read in so many deindustrial studies texts. As one interviewee summarised the situation: 'Grangemouth has got all the symptoms of a town that industries have left.' In 2025, INEOS effectively closed Grangemouth's refinery due to profitability concerns (while the petrochemical plants remain), in a completely unjust 'transition'.[8] Back in 2019, however, the interviewee's remark stuck in my mind as counterintuitive, even absurd, for a place where smokestack fumes are visible throughout the day, an orange glow shines all night and the smell of chemicals is familiar to all inhabitants, along with the flames and vibrations produced by heavy flaring (the combustion of excess gases and liquids through flare stacks). I only had to look out of my B&B window to see the flashing lights and smoke clouds emanating from the petrochemical complex. And yet, the retired petrochemical worker just quoted was referring to a very real phenomenon.

To make sense of this paradox, I went back to three bodies of literature: political economy, deindustrial studies and environmental justice studies. In political economy, deindustrialisation most often refers to declining shares of industrial jobs in total employment, regardless of output.[9] In her critical review, Fiona Tregenna notes that 'deindustrialisation is typically conceptualised as a decline in manufacturing as a share

7 Alessandro Portelli, *Biography of an Industrial Town: Terni, Italy, 1831–2014*, London: Palgrave Macmillan, 2014, 374.

8 Riyoko Shibe and Ewan Gibbs, 'Workers' Perspectives on an Unjust Transition: Place, History, and Workplace Closure at Grangemouth Oil Refinery, Scotland', *Antipode* 57(4), 2025, 1576–97.

9 See, e.g., Robert Rowthorn and Ramana Ramaswamy, 'Deindustrialization: Causes and Implications', IMF Working Papers 97/42, 1997.

of total employment'.[10] Thus defined, deindustrialisation can easily co-exist with continuing or even growing industrial production, particularly when productivity-raising technologies are being introduced.

Deindustrial studies has a different understanding of 'deindustrialisation'. In this literature the term does not normally refer to declining industrial employment shares, but to factories shutting down. Deindustrial studies ventures beyond the statistics to foreground the human costs of deindustrialisation, recounting the lived experiences of workers and industrial communities facing plant closures.[11] The unemployment, precarity, fraying of community networks and loss of purpose that comes with deindustrialisation has also been portrayed in many UK deindustrialisation movies, from *The Full Monty* to *Brassed Off*.

By contrast, the main focus of environmental justice studies is the uneven distribution of burdens caused by noxious industries.[12] Therefore, while deindustrial studies concentrates on the sufferings brought to communities by factory closures (e.g. unemployment and weakening community ties), environmental justice studies conversely foregrounds the strains generated by operating factories (e.g. environmental degradation and health effects). These two sets of problems are usually seen as mutually exclusive, but if deindustrialisation is understood in terms of declining employment shares, rather than as plants shutting down, then both can be visited upon the same populations simultaneously. In reality, underemployment and smog are far from incompatible.

At first glance, noxious deindustrialisation could be presumed to be a niche and even transient phenomenon, specific to those former 'white working class' strongholds in the Global North where large-scale industrial production still lingers on. So far, indeed, deindustrial studies

10 Fiona Tregenna, 'Characterising Deindustrialisation: An Analysis of Changes in Manufacturing Employment and Output Internationally', *Cambridge Journal of Economics* 33, 2009, 433–66, 433.

11 See Barry Bluestone and Bennett Harrison, *The Deindustrialization of America: Plant Closings, Community Abandonment, and the Dismantling of Basic Industry*, New York: Basic Books, 1984.

12 See Dorceta E. Taylor, *Toxic Communities: Environmental Racism, Industrial Pollution, and Residential Mobility*, New York: New York University Press, 2014.

mostly appears to have taken a standpoint limited to the Global North working class as opposed to an internationalist one, seldom questioning the utter dominance of industrial exports enjoyed by the Global North up until the beginning of 'globalisation'.[13] As Schindler and collaborators note: 'The Global South looms large in the background of deindustrialization scholarship, but it remains a non-descript destination for runaway factories. The implicit assumption embedded in much of this scholarship is that the South has industrialized at the expense of cities and communities in the North Atlantic.'[14] On the one hand, this assumption overlooks deep differentiations within the Global South, as well as how deindustrialisation has been more severe in Africa and South America in terms of missed 'development opportunities' relative to the already highly industrialised Global North. On the other hand, it is clear that calls for a 'reindustrialisation' of the Global North are hardly compatible with a more balanced international division of labour. The following pages stress the *global* dimension of noxious deindustrialisation as driven by the imperatives of capitalist accumulation.

The prime mover of the ecological crisis

Since employment deindustrialisation is measured by the share of industrial jobs in total employment, within a given area, noxious deindustrialisation is a scalable concept: it can be applied to local communities as well as at national, regional and global levels. This section provides a theoretical framing to understand global noxious deindustrialisation, before the following sections offer an empirical overview of the phenomenon. For now, suffice it to say that, on the one hand, the global share of industrial employment has slowly declined over recent decades; on the other hand, one hardly needs to demonstrate the other face of global

13 This is currently changing, particularly through the Deindustrialization and the Politics of Our Time (DEPOT) research project.

14 Seth Schindler et al., 'Deindustrialization in Cities of the Global South', *Area Development and Policy* 5(3), 2020, 283–304, 285.

noxious deindustrialisation, that is, the progressive accumulation of noxiousness on a planetary scale. A vast amount of scientific evidence has already proven that we are increasingly moving past the planetary boundaries that guarantee the safe reproduction of life on the planet.[15]

Global noxious deindustrialisation can be interpreted as an outcome of Marx's 'general law of capitalist accumulation', according to which capitalist technological development tends to augment labour productivity by raising the organic composition of capital, which, in turn, engenders an increase of employment precarity.[16] As summarised by Marx, 'the higher the productivity of labour, the greater is the pressure of the workers on the means of employment, [and] the more precarious therefore becomes the condition for their existence'.[17] The general law of capitalist accumulation has then two dialectically interrelated faces. It features the co-evolution of both *capital composition*, seen in the rising organic composition of capital, and *class composition*, seen in the rising share of the surplus working class. This chapter is centred on capital composition, while the next one focuses on class composition.

The first thing to note is that what Marx called a 'law' can be offset by an array of more or less contingent countertendencies. The most obvious countertendency is output expansions large enough to compensate for the underemployment induced by productivity gains and the enlargement of the working class. However, as discussed below, the in-built propensity of capitalism for crises of overproduction makes it unlikely that output expansions will be big and durable enough to fully neutralise the tendency towards rising precarity. The latter is then a constant force in capitalism, the effects of which become evident in the absence of powerful enough pushbacks.

In Marx's terminology, capital has a technical composition, a value composition and an organic composition, all of which point to the relationship between workers and the means of production under

15 Katherine Richardson et al., 'Earth Beyond Six of Nine Planetary Boundaries', *Science Advances* 9, 2023, 2458.

16 Karl Marx, *Capital: A Critique of Political Economy*, vol. 1, London: Penguin, 1976 [1867], 762–870.

17 Ibid., 798.

capitalism. Technical composition refers to the relation between the physical means of production (buildings, machines, raw materials, energy, etc., which Marx calls 'dead labour' because they embody the past work of the people who produced them) and the workers (the 'living labour') who set the means of production in motion. Value composition designates the proportion between the cost of the means of production (which Marx calls 'constant capital') and the wage bill (or 'variable capital'). According to Marx, the technical and value compositions of capital tend to augment in tandem, since the productivity-raising technological development of the means of production results in a rising share of constant capital over variable capital. The organic composition of capital then expresses the co-evolution of its technical and value compositions; it is 'the value composition of capital, in so far as it is determined by its technical composition and mirrors the changes in the latter'.[18] For example, the latest generation extreme ultraviolet (EUV) lithography machine, necessary for the printing of microchip transistors on silicon wafers, costs more than 350 million euros.[19] This expensive machinery requires vast amounts of electricity and water to operate but only a limited number of workers, meaning that advanced semiconductor production is a capital-intensive branch. In Marx's terms, it has a high organic composition of capital.

Capital is propelled by horizontal and vertical conflicts (intra-class market competition and inter-class struggle) to raise labour productivity in order to keep up with rival firms and hold workers' offensives in check. As seen in the introduction, initially, a productivity-raising technology is held exclusively by cutting-edge firms, which – by monopolising it – reap surplus profits relative to their competitors. For example, the Dutch corporation Advanced Semiconductor Materials Lithography (ASML) is currently the only company in the world producing EUV lithography machines, which grants it a profit beyond the 'normal' rate. However, technological innovations then spread within and across sectors,

18 Ibid., 762.

19 Tim Bradshaw and Anna Gross, 'The Big Question of How Small Chips Can Get', *Financial Times*, 31 May 2023.

equalising costs, expelling workers, restoring competition, and prompting capital to innovate again.

Productivity gains eventually cheapen wage goods (that is, products satisfying workers' consumption needs), which, in turn, lowers the value of labour power, since it takes less average labour time to produce the same amount of means of reproduction. This allows capital to appropriate an increasing share of the product relative to labour, a phenomenon that Marx called the accumulation of 'relative surplus value'.[20] For example, the 'modernisation' of agriculture through Green Revolution technologies (machines, chemical fertilisers, pesticides and high-yielding seeds) had devastating effects on peasants' livelihoods and on biodiversity.[21] At the same time, however, it compressed the average labour time needed to produce given amounts of food. With cheaper food, if *real* wages stay the same (if workers continue to buy similar quantities of nutrients), then *relative* wages (the wage share in output) will decline. Conversely, the rate of exploitation (the proportion between surplus value and the wage bill or variable capital) will increase. *Relative* wage cuts, then, do not necessarily translate into *real* wage cuts. In fact, if the combination of class struggles and capital's need for a more qualified workforce improves standards of living, relative wage cuts become compatible with real wage rises. Nonetheless, as the reproduction of the working class is cheapened by technological development, so the share of the total social product appropriated by capital tends to increase in relation to that appropriated by labour; the labour share of income tends to decline and the profit share of income tends to rise.

The relationship between the general law of capitalist accumulation and employment deindustrialisation has been highlighted by the journal *Endnotes*, and in particular by Aaron Benanav.[22] According to Benanav, global employment deindustrialisation was not brought about by the prowess of technological innovation per se, as is claimed

20 Marx, *Capital*, vol. 1, 429–38.

21 Vandana Shiva, *The Violence of Green Revolution: Third World Agriculture, Ecology and Politics*, Lexington: University Press of Kentucky, 2016 [1989].

22 Aaron Benanav and John Clegg, 'Misery and Debt', *Endnotes* 2, 2010, 20–51; Aaron Benanav, *Automation and the Future of Work*, London: Verso, 2020.

by a wide range of commentators on automation.[23] In principle, *industrial output expansions that are large enough to compensate for productivity gains and for the demographic growth of the working class would have the effect of maintaining unaltered the share of the industrially employed working class.* Imagine automation cuts the time needed to make a smartphone by half, while the working population doubles: if smartphone output remains the same, the share of smartphone producing jobs in total employment will become one fourth of what it used to be. However, if smartphone output quadruples, this share will remain unaltered. Benanav has shown that profit-driven technological change has led to global employment deindustrialisation and growing precarity in recent decades not because of exceptionally high productivity gains, but because of exceptionally low output growth. In fact, since the 1970s, productivity has increased more slowly than in the post–Second World War decades, and gross domestic product (GDP) has grown even slower. In our example, smartphone output has risen by a factor inferior to four, causing a decline in the share of workers employed by smartphone factories.

As productivity gains in the service sector are even slower than in industry, stagnation in industry translates into overall economic stagnation. Concurrently, the slow pace of productivity gains in the service sector relative to industry ensures that the workforce that is expelled from the latter ends up in the former. In our example, automation kicks workers out of smartphone factories. Since customer assistance jobs are harder to automate, because consumers still prefer dealing with humans rather than with robots, the share of customer assistance jobs in total employment increases, as workers exit factories to enter call centres. In a context of stagnation and dwindling labour income shares, however, most new employment created in the service sector tends to be precarious. Because most people still need to work for a living, the outcome of automation-cum-stagnation is not apocalyptic joblessness but, as we will

23 See, e.g., Erik Brynjolfsson and Andrew McAfee, *The Second Machine Age: Work, Progress, and Prosperity in a Time of Brilliant Technologies*, New York: W. W. Norton and Company, 2014.

see in the next chapter, precarity rising where job security was high and precarity persisting where it has always been prevalent.[24]

Capitalist 'creative destruction' is not an eternal return of the same. Rather, it tends towards increasing job insecurity and falling relative wages.[25] This can be seen empirically in the evaporation of the Fordist promise of a generalisation of 'standard' employment and in the decline of the labour share of income over the last decades.[26] Countries that never reached high industrialisation levels 'import' deindustrialisation together with cheap manufactured products or advanced industrial machinery.[27] In the latter case, output industrialisation can well coincide with employment deindustrialisation. But *why* does industrial output not rise fast enough to compensate for productivity gains and the demographic expansion of the working class, so as to keep the share of industrial employment constant?

The rise in the organic composition of capital in industry means that dead labour – the past labour accumulated in the means of production on an ever-larger scale – weighs more and more upon living labour, and thus upon the production of new value. Fewer workers are necessary to set in motion more sophisticated machineries and less worktime is needed to make growing quantities of commodities. The wage bill that accrues to living labour constitutes a shrinking share of total costs, while constant capital, the embodiment of dead labour monopolised and personified by the capitalist class, tends to expand its share. Higher productivity increases material output, but if this higher material output does not meet adequate levels of demand, prices can fall to such an extent that profitability is strained. Beyond a certain point, the race to

24 Aaron Benanav, 'Precarity Rising', *Viewpoint Magazine*, 15 June 2015, viewpointmag.com.

25 See Alexis Moraitis, 'From the Post-Industrial Prophecy to the De-Industrial Nightmare: Stagnation, the Manufacturing Fetish and the Limits of Capitalist Wealth', *Competition and Change* 26(5), 2022, 513–32.

26 International Labour Organization, *Global Wage Report 2022–23: The Impact of Inflation and COVID-19 on Wages and Purchasing Power*, Geneva: International Labour Office, 2022.

27 Dani Rodrik, 'Premature Deindustrialization', *Journal of Economic Growth* 21(1), 2016, 1–33.

produce more in less time generates a full-blown crisis of overproduction. Investment rates crash and output growth is held back.

Individual firms have two main ways to counteract low profitability: raising the rate of exploitation further and cheapening constant capital. Employment precarity is functional to the first countertendency. In fact, the expansion of the surplus working class affords the intensification of labour while containing wages, because workers who do not accept such conditions can be more easily substituted with members of the 'reserve army of labour'. Environmental degradation is necessary to the second countertendency, as the appropriation of nature on an ever-increasing scale is a condition for lowering the value of capital goods. In this sense, employment precarity and environmental degradation are both manifestations of the capitalist imperative to cut costs in order to keep the economy moving.

However, while these countertendencies have successfully prevented the apocalyptic collapse of capitalism expected by so many leftists, they come with their own contradictions. On the one hand, the limit to all strategies to raise the rate of exploitation is constituted by capital's need for growing demand in order to turn a profit at all.[28] On the other hand, as 'the value of the constant capital does not increase in the same proportion as its material volume', the drive to convert more and more of nature into capital runs into the material limits of the earth's capacity for regeneration.[29] This hinders progress in cheapening capital goods.[30]

To give an example, the cheapening of smartphones (which are both capital and wage goods) requires low-cost labour and huge quantities of minerals, plastics, energy and water, as well as the discharge of emissions and waste into the environment.[31] A growing market for smartphones is

28 Karl Marx, *Capital: A Critique of Political Economy*, vol. 3, London: Penguin, 1991 [1894], 364–8.

29 Ibid., 343.

30 See James O'Connor, 'Capitalism, Nature, Socialism: A Theoretical Introduction', *Capitalism, Nature, Socialism* 1, 1986, 11–38.

31 Jenny Chan, Mark Selden and Pun Ngai, *Dying for an iPhone: Apple, Foxconn, and the Lives of China's Workers*, London: Pluto Press, 2020; James Suckling and Jacquetta Lee, 'Redefining Scope: The True Environmental Impact of Smartphones?', *International Journal of Life Cycle Assessment* 20, 2015, 1181–96.

kept alive by expanding credit to the working class, making internet access necessary to workers in all corners of the planet, keeping the useful life of devices short and their reparation difficult, and marketing a desire for new trendy models. Yet, if the rate of exploitation rises too quickly, workers will ultimately not have enough money to keep the demand for smartphones in pace with growing productive capacity, and another crisis of overproduction will follow. At the same time, as seen with the chip shortage of the early 2020s, raw materials cannot always be brought online at the pace required by the market, which curtails or delays output expansion.

In sum, the contradiction between capital's need to contain costs and the necessity to find buyers for its products, combined with bottlenecks in the appropriation of nature, has kept industrial output growth in check, ensuring that in recent decades it did not expand fast enough to compensate for productivity gains in industry and prevent slow global employment deindustrialisation. In turn, slow growth in the service sector, combined with the continued dispossession of the peasantry, has ensured that employment deindustrialisation contributed to an increase of employment precarity across sectors, as shown in the next chapter.

The evolution of capital composition with its tendency towards increasing dead labour over living labour is not, however, just a quantitative phenomenon. It is also a qualitative one. Automation not only spurs a *quantitative* increase in the proportion of constant capital over variable capital but also engenders a *qualitative* tightening of capitalist domination over the labour process and the workers engaged in it. As Panzieri observed in 1961:

> *Capitalist* technological development involves, through different phases of rationalisation and evermore refined forms of integration, a continual growth of capitalist control. The fundamental factor in such a process is the growing rise of constant capital over variable capital . . .
>
> The progressive widening of the 'social abyss' between workers and capitalists was also expressed by Marx through the form of the *relative*

> *wage* as well as its decline. Yet, it is clear that such a concept implies an element of political consciousness, that is, the awareness that the improvement of material conditions – the rise of the nominal and real wage – comes with a deepening of 'political dependence'.[32]

In this sense, one may speak of the 'rising *political composition of capital*' to refer to capital's drive – always resisted to some extent – to secure its power over the labour process.

This qualitative dimension of constant capital, distinct from the quantitative aspect of *how much* technology is deployed in the labour process, concerns which technologies are developed. That is, the *how* and *what* of production. As highlighted by what Andrew Feenberg calls 'critical theories of technology', there are many possible trajectories for technological development, making the political composition of capital itself a terrain of struggle.[33] André Gorz wrote: 'Without the struggle for a different technology, the struggle for a different society is in vain.'[34] Or, in Panzieri's words: 'Working-class overthrow of the system is a negation of the entire organisation in which capitalist development is expressed – and first and foremost of [capitalist] technology insofar as it is linked to productivity.'[35]

Class struggle is far from being the only social force acting upon technological development, but it has an important (and contradictory) role in both limiting and driving it. In fact, technological innovation in response to class struggle has two aspects: concession and outmanoeuvring. If technological innovation merely occurred to accommodate the 'critique of plant engineering from a working-class perspective' (for example, via 'prevention-through-design' engineering exclusively geared towards maximising workers' well-being and

32 Raniero Panzieri, 'The Capitalist Use of Machinery: Marx Versus the "Objectivists"' [1961], in Phil Slater (ed.), *Outlines of a Critique of Technology*, Atlantic Highlands, NJ: Humanities Press, 1980, 39–68, 48–9, 60.

33 Andrew Feenberg, *Transforming Technology: A Critical Theory Revisited*, Oxford: Oxford University Press, 2002.

34 André Gorz, *Ecology as Politics*, Boston: South End Press, 1979 [1978], 19.

35 Panzieri, 'The Capitalist Use of Machinery', 60.

control), then the competitiveness of firms adopting it would be compromised.[36] To ensure that firms incorporating noxiousness-preventive innovations remain competitive, technological change also requires the reassertion of capitalist discipline at ever higher levels of development. For example, latest generation assembly lines are designed so as to minimise ergonomic stress to the workers. However, they also incorporate sophisticated digital surveillance technologies that allow the pace of work to be intensified, which is often necessary to recover the cost of buying them. Needless to say, the intensification of assembly line work brings noxiousness back in, albeit in a new form. In other words, *actually existing capitalist technologies are a contradictory, provisional synthesis of the antagonistic duality of concession and outmanoeuvring*. Nonetheless, they must always be compatible with the profit imperative in order to become generalised. The political composition of capital thus bears the imprint of both the law of value and workers' antagonism towards it.

Of course, this reasoning applies only at a certain level of complexity. There is no doubt a place for hammers, copper cables and, indeed, microchips in an alternative mode of production. As Panzieri wrote, 'the parts of the labour process compatible with social regulation must be critically extricated from the capitalist nexus of technology and power'.[37] These elements would thus be produced and deployed through different complex technological systems relative to the ones being developed today.

Examples of the dialectic between class struggle and technological innovation are manifold, from the self-acting spinning mule to the Taylorist–Fordist assembly line, numerically controlled machinery, containerisation and, more recently, the 'algorithmic management' made possible by digital tracking and big data analytics.[38] Yet, in relation to the ecological crisis, the most significant area of development is

36 Sergio Bologna, 'La tribù delle talpe', *Primo Maggio* 8, 1977, 3–18, 11.

37 Panzieri, 'The Capitalist Use of Machinery', 20.

38 Phoebe V. Moore, *The Quantified Self in Precarity: Work, Technology and What Counts*, Abingdon-on-Thames: Routledge, 2018.

energy.[39] According to Andreas Malm, coal replaced water streams as the main industrial energy power source due to capital's need for mobility in its quest for cheap labour. This occurred in Britain in the 1830s, during a period of rising labour unrest whereby 'the struggle against labour called for machinery, which called for steam power, which called for coal, thereby coupled to the growth of manufacturing'.[40]

Timothy Mitchell similarly interprets the (partial) substitution of coal with the more fluid substances of oil and gas as a technological development geared towards neutralising workers' 'power of inhibition'.[41] This was achieved through the minimisation of choke points in the energy flow with the pipeline acting as a device for circumventing working-class intransigence. The slippery nature of the pipeline was also noted by the Porto Marghera *operaista* group, in their analysis of the technological restructuring that followed the 1968–69 strike wave at the petrochemical complex known as Petrolchimico. In 1970, a new ethylene steam cracker was installed in Porto Marghera, which inaugurated 'a disconnection between productive expansion and employment expansion'.[42] The new Petrolchimico was integrated via a pipeline network to the petrochemical hubs of Ferrara and Mantova. The workers' collective commented: 'Those pipelines present themselves as innocuous, but they are actually highways taking away what we produce before our eyes.'[43]

Just like capitalist technological innovation, the noxious deindustrialisation resulting from this development is not a linear and neutral process, but rather one driven by antagonism. The tendency of the organic composition of capital to rise – both in its quantitative, economic dimension and in its qualitative, political one – leads to the expulsion of an increasing share of workers from industry, through the adoption of

39 See Davide Gallo Lassere, 'Welcome to the Past: Autonomy of Nature, Fossil Fuels and the Capitalocene', Verso Books blog, 24 June 2022, versobooks.com.

40 Andreas Malm, *Fossil Capital: The Rise of Steam Power and the Roots of Global Warming*, London: Verso, 2016, 222.

41 Timothy Mitchell, *Carbon Democracy: Political Power in the Age of Oil*, London: Verso, 2011, 108.

42 *Potere Operaio*, November 1973, 13.

43 *Controlavoro*, 25 April 1977, 3.

technologies that ensure capital's control over the labour process and deepen environmental degradation. Noxious deindustrialisation was thus well captured, *ante litteram*, in this passage from Harry Braverman:

> In the automobile industry, a constantly diminishing number of workers produces, decade by decade, a growing number of increasingly degraded products which, as they are placed upon the streets and highways, poison and disrupt the entire social atmosphere – while at the same time the cities where motor vehicles are produced become centers of degraded labor on the one hand and permanent unemployment on the other.[44]

Paraphrasing Aristotle and his notion of 'prime mover' (known in Italian as '*motore immobile*'), Tronti cast the working class as the 'moving mover [*motore mobile*] of capital'.[45] In this sense, class struggle is the prime mover of capitalist development, functioning at a much deeper level than the *dead fossil-fuel power* resurrected to confine *living working-class power* within the parameters of capital accumulation. No doubt Tronti, writing in 1965, did not have the ecological crisis in mind. Today, however, we can see how working-class power, driving the development of technologies capable of controlling it, has become, in perverted form, the fuel in the engine of the ecological crisis. The fire of class struggle powers capitalist technological development, which sets the world ablaze in turn. And yet, in class struggle lies manifold possibilities for reforming capitalist technology, as well as the hope of reaching the 'off' switch, laying the groundwork for the production of wealth through alternative social relations and thus alternative technologies.

44 Harry Braverman, *Labor and Monopoly Capital: The Degradation of Work in the Twentieth Century*, New York: Monthly Review Press, 1974, 141.

45 Mario Tronti, *Workers and Capital*, London: Verso, 2019 [1966], 204.

Deindustrial decline with industrial noxiousness

The secular tendency of the global share of primary sector employment to fall is well established.[46] The peasants, fishers and herders of the world – together with their proletarianised colleagues in the same sectors – are still a significant, but nonetheless shrinking, portion of the global workforce. However, the global share of industrial jobs in total employment, as measured by the ILO under the label of 'manufacturing employment', has also gradually been declining. In fact, it fell from 15.7 per cent in 1991 (the first year for which such estimates are available) to 13.6 per cent in 2021, while employment shares in construction and services rose.[47] These estimates should be taken with a grain of salt, as they are periodically updated with slightly different numbers each time. Nonetheless, a slow decline in industry's global employment share over the past three decades has been reported consistently. While in modernist imaginaries the prominence of industrial work appeared to be the future of (mostly male) humanity, today it looks more like a historical parenthesis, propelling a transition from agricultural to service sector employment.

Part of this decline in the global share of industrial employment stems from productivity gains in industry, which are more rapid than both productivity gains in other sectors and industrial output growth. As the United Nations Industrial Development Organization (UNIDO) summarises: 'the world [real] manufacturing value added grew faster than the total economy. But world manufacturing employment grew slower than total world employment. So, labour productivity in the manufacturing sector grew faster than in the total economy.'[48] An additional factor is the outsourcing of previously in-house jobs from

46 Farshad A. Araghi, 'Global Depeasantization, 1945–1990', *Sociological Quarterly* 36(2), 1995, 337–68.

47 Unless otherwise stated, the figures presented in this section are calculated based on the ILOSTAT Modelled Estimates repository and the UNIDO National Accounts Database.

48 UNIDO, *Industrial Development Report 2020: Industrializing in the Digital Age*, Vienna: UNIDO, 2019, 150.

industrial to service sector firms (in logistics, security, cleaning, catering, etc.). The latter phenomenon, however, is not a mere change of labels because outsourced jobs are, on average, more precarious and lower paid than in-house ones, which is in line with Marx's general law of capitalist accumulation (which, remember, states that capitalist development tends towards higher capital intensity as well as higher precarity).

Of course, global employment deindustrialisation does not mean that industrial production itself has waned. To explore this issue, let us start with the difference between *nominal* and *real* manufacturing value added (MVA). Nominal MVA takes the evolution of prices in time at face value. Real MVA, conversely, holds prices constant at a given base year. Imagine a two-sector economy made of smartphone factories and barber shops. Over one decade, new technologies are introduced in smartphone factories, so that making one smartphone takes one tenth of the time it used to. The haircutting labour process, however, remains fairly close to what it has been for centuries, meaning that giving a haircut still takes the same time. As a consequence, the price of smartphones will decline relative to the price of haircuts, ensuring that the share of nominal MVA in GDP will shrink unless smartphone output increases extremely quickly. The share of real MVA in GDP, however, will increase as the prices of Year 1 are held constant, reflecting the fact that the number of smartphones produced has grown faster than the number of haircuts given.

According to UNIDO estimates, the global share of *nominal* MVA in total value added has declined from 20.3 per cent in 1990 to 15.3 per cent in 2009 and has since then stagnated between 15 and 16 per cent. This is because more rapid productivity gains in manufacturing, relative to other sectors, have ensured that inflation has raised the nominal prices of manufactured goods to a lesser extent than those of other products. Conversely, *real* MVA has grown more rapidly than value added in other sectors, but for the very same reason: faster productivity gains in the former than in the latter. Therefore, the global share of real MVA in GDP (in constant 2015 dollars) has risen from 14.7 per cent in 1990 to 16.7 per cent in 2020. The fall in industrial nominal price output (relative to total value added) coupled with the (relative and absolute) increase of industrial material output thus indicates how the continuous shortening of the average labour time needed to create wealth is particularly rapid in industry.

The decline in the global share, or *rate*, of the industrially employed workforce is then perfectly compatible with – and has been accompanied by – both growing industrial material output and a rising absolute number, or *mass*, of industrial workers. Rates are relative quantities, masses are absolute ones; this distinction is crucial in Marx. In 1991, the factory workers of the world were 350 million; by 2021 their number was 445.5 million, the largest mass of industrial workers in human history up to that point. The fall of industrial employment is thus *relative*, both to the magnitude of constant capital it sets in motion (and thus to the amount of material goods it produces) and to the size of the working class as a whole. In other words, *employment deindustrialisation is relative to the total capital and class compositions.*

While paradoxical, this trend simply parallels the broader, double-edged decline in the rate of living labour over dead labour (that is, of variable capital over constant capital), accompanied by growth in the mass of living labour itself. Marx noted that 'the relative decline in the variable capital and profit goes together with an absolute increase in both', and added that bourgeois economists invoked the rising mass of profit as a kind of consolation in cases of its falling rate.[49] One could argue that some socialist economists are similarly prey to wishful thinking in regards to the 'progressive' direction of capitalist development today, as they invoke the rising mass of the industrial workforce as a kind of consolation for the falling rate of industrial employment.

The fact that productivity growth is faster in manufacturing than in other sectors says less about its speed and more about the waning dynamism of the economy as a whole. The latest drivers of productivity gains are Advanced Digital Production (ADP) technologies, defined by UNIDO as a 'combination of hardware (advanced robots and 3D printers), software (big data analytics, cloud computing and artificial

49 Marx, *Capital*, vol. 3, 329, 330.

intelligence) and connectivity (the internet of things)'.[50] The connectivity between hardware and software is achieved through the incorporation of digitally connected sensors and actuators in the machinery, which is then able to collect and transmit data to the software and react according to the latter's instructions. The cutting-edge microchip production I have been using as a source of examples is key to these innovations. In combination with other elements such as nanotechnologies, biotechnologies and – crucially – advanced renewable energy, these digital networks (or 'cyber-physical systems') constitute what is known as 'Industry 4.0'. Such developments follow the 'third industrial revolution', which began in the 1960s with the gradual introduction of microchips and computers. Yet, despite both utopian expectations of gargantuan productivity gains and dystopian concerns over apocalyptic unemployment rates, the technologies of Industry 4.0 have not boosted productivity as rapidly as previous waves of innovation.[51] To the contrary, hefty productivity gains have remained concentrated in leading high-tech companies. Failing to diffuse across sectors and countries, these technologies have not accelerated global productivity growth as a whole.[52]

Slowing productivity growth has affected the Global North since the 1970s.[53] At the global level, this has been compensated for by rapid productivity gains in 'developing' countries, but only up until the 2007–08 financial crisis, after which productivity growth has also been decelerating in the Global South.[54] Stagnation is thus increasingly becoming a global reality. One cannot help noticing that this trend began with the defeat of the 'Long 1968' cycle of struggles, in which dwindling profit

50 UNIDO, *Industrial Development Report 2022: The Future of Industrialization in a Post-Pandemic World*, Vienna: UNIDO, 2021, xvi.

51 See Klaus Schwab, *The Fourth Industrial Revolution*, Cologny: World Economic Forum, 2016.

52 David Autor et al., 'The Fall of the Labor Share and the Rise of Superstar Firms', *Quarterly Journal of Economics* 135(2), 2020, 645–709.

53 Robert J. Gordon and Hassan Sayed, 'The Industry Anatomy of the Transatlantic Productivity Growth Slowdown: Europe Chasing the American Frontier', *International Productivity Monitor* 37, 2019, 3–38.

54 Alistair Dieppe (ed.), *Global Productivity: Trends, Drivers, and Policies*, Washington, DC: World Bank, 2021.

rates were met with a radicalisation of labour militancy in the Global North and national liberation struggles in the Global South. Profitability was restored by fragmenting industrial production and increasingly dispersing it from the Global North to parts of the Global South.

Yet, as Tronti noted, repression carries a cost for capital, bringing about a decline in the dynamism of its own accumulation:

> When the capitalists call the working class into an open confrontation, in order to defeat its political movement on the battlefield, they then pay for their momentary success with the long periods of passivity that living labour introduces, in response, into the economic mechanism.[55]

In this sense, the scant diffusion of Industry 4.0 technologies could be interpreted as the result of a relative weakening of the incentive constituted by successful active workplace struggles. Working-class passivity thus becomes capitalist passivity in the form of the much-bemoaned weak investment rates of our time: 'Originating in the aftermath of major recessions in OECD economies, slow investment growth has become a concern for many other regions in the world and is most pronounced in the largest emerging markets and in commodity exporters.'[56] Stagnation is, therefore, the price capital pays for repression, even though capital needs repression to defend profitability.

Coming now to the 'noxious' in noxious deindustrialisation, global employment deindustrialisation is occurring in a context of cumulative production-driven environmental degradation. Let us take CO_2 emissions. Admittedly, they are a simplistic proxy for noxiousness. This is because, as the 'planetary boundaries' approach shows, the ecological crisis is much more than climate change.[57] Its other dimensions are biosphere integrity (biodiversity loss), land-system change (deforestation), freshwater use (pressures on the hydrological cycle), biogeochemical flows

55 Tronti, *Workers and Capital*, 316.

56 International Labour Organization, *World Employment and Social Outlook: Trends 2023*, Geneva: International Labour Office, 2023, 97.

57 Richardson et al., 'Earth Beyond Six of Nine Planetary Boundaries'.

(disruptions of the nitrogen and phosphorus cycles), ocean acidification, atmospheric aerosol loading (air pollution), stratospheric ozone depletion (the 'ozone hole' made by artificial chlorine-containing gases) and novel entities (chemical pollution). This means that even if net-zero CO_2 emissions were reached thanks to a successful energy transition, the ecological crisis would continue due to the persistence of further imbalances induced by capitalist noxiousness.

With this caveat in mind, tracking CO_2 emissions is still nevertheless helpful for evidencing noxious deindustrialisation. Figure 1.1 shows how the declining share of global industrial employment has not led to a decline in CO_2 emissions, whether released by industrial production directly or by the deployment of industrial products in other sectors and in final consumption. In fact, exactly the opposite has happened.

Figure 1.1 Trend comparison of global manufacturing employment (%) and CO_2 emissions (base year: 1991 = 100, 1991–2021)

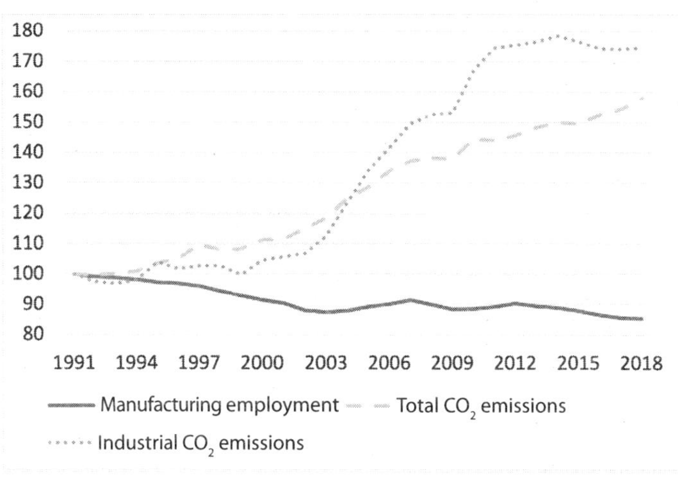

Source: ILOSTAT and Climate Analysis Indicators Tool.

Since the dawn of capitalism, the global share of industrial employment and CO_2 emissions had long increased in tandem. However, as Figure 1.1 indicates, at some point in the second half of the twentieth century, the two historical series decoupled: the share of industrial employment peaked and began to fall, while, despite this decline, yearly CO_2 emissions continued to soar. So far, rather than a decoupling between GDP growth and excessive use of material resources, we see a disconnection between industrial production and industrial employment, while the unsustainable appropriation of nature continues.

Furthermore, emissions have kept increasing in spite of improvements in the environmental performance of industry. Global energy efficiency in industry has been growing since 2010, meaning that the sector's energy intensity (energy consumption per unit of MVA) decreased by 15 per cent between 2000 and 2018.[58] According to International Energy Agency (IEA) data, the energy intensity of the global economy as a whole has also diminished significantly, falling by 36.5 per cent between 1990 and 2020. In a glaring illustration of the Jevons Paradox we encountered in the introduction, the unabated growth in CO_2 emissions is entirely explained by the continuous rise in global material output. In a way, then, trends in CO_2 emissions and industrial employment have evolved in parallel: in both cases their mass (i.e. absolute quantity) has increased, while their rate to output (i.e. their relative quantity) has declined. Unfortunately, nonetheless, the only trend that matters for climate change is the rise in the *absolute* quantity of CO_2 emissions.

Noxious deindustrialisation is a global trend, but it manifests itself in extremely different ways across countries, depending on the timing and scale of national-level industrialisation. For example, in an early industrialiser such as the US, the trend towards employment deindustrialisation had already begun in the 1950s. In a late industrialiser like Italy, employment deindustrialisation only started in the 1980s. In a 'very late' industrialiser such as China, employment deindustrialisation has only surfaced in the past decade. In countries whose industrialisation attempts never really came to fruition, for example Morocco, employment in manufacturing peaked before the country can be said to have industrialised at all. This can be visualised in Figure 1.2, where the vertical bars represent national manufacturing employment shares and the rising lines show CO_2 emissions.

58 UNIDO, *Industrial Development Report 2022*, 20.

Figure 1.2 The unevenness of noxious deindustrialisation

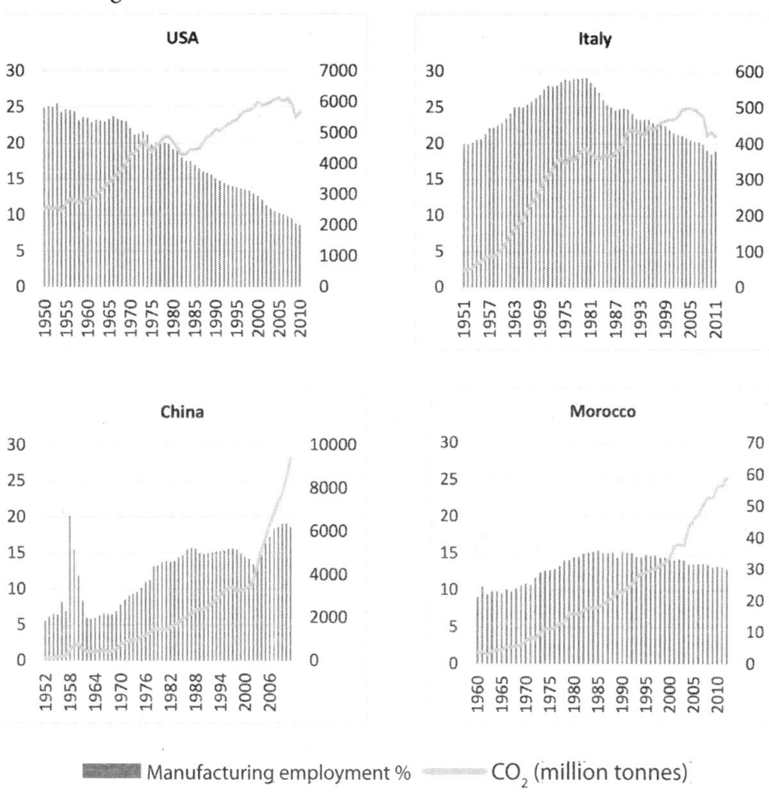

Source: Calculated by the author based on GGDC 10-Sector Database and Global Carbon Project.

Such different pathways reflect an international division of labour that has undergone radical transformation over time, but has thus far remained dominated by the Global North. In 1990 (the year for which UNIDO estimates begin), about 80 per cent of the global MVA was still concentrated in North America, Europe and Japan, regions which at the time hosted only 21 per cent of the world's population. In the following three decades, however, the Global North shed factory jobs fast: industry as a share of total employment within North America, Europe and Japan declined from 21.4 per cent in 1991 to 13.5 per cent in 2021. Nevertheless, in 2020, these countries still accounted for over 47 per cent of the global share of MVA, while hosting only 16 per cent of the world's population.

As a result of restructuring in the Global North, industry expanded massively in Asia. Between 1990 and 2020, Asia's share of the world's MVA (excluding Japan) expanded from a mere 10.8 to 45.5 per cent. However, this process has been highly uneven within Asia itself. As is well known, the largest player by far is China, which has seen its share of the world's MVA skyrocket from 2.1 per cent in 1990 to an astounding 28.6 per cent in 2020. However, industry as a share of total national employment peaked in China in the early 2010s at around 22 per cent and was estimated to have declined to 19 per cent in 2020.

Industry has, in contrast, been contracting (in relative terms) in Africa and South America. Together with the oil exporting countries of the Middle East, these regions have mainly been returned to their classic role in the international division of labour as providers of raw or 'quasi-raw' materials. The defeat of twentieth-century attempts at import-substitution industrialisation in these regions, which might have resulted in a more even distribution of technological capabilities worldwide, has been dubbed 'reprimarisation', as it marks a return to primary sector dominance.[59] The *combined* share of the world's MVA in Africa and South America declined from 5.7 per cent in 1990 to 4.5 per cent in 2020. This has impacted on employment, with industry's share in total employment falling from 8.2 per cent in 1991 to 7.4 per cent in 2021 in Africa and from 15 to 11 per cent for South America.

The overall (and still highly simplified) picture is therefore one in which the Global North continues to dominate the technological frontier, and thus high-productivity industry, while large parts of Asia compete for labour intensive, low-wage industry, and the Middle East, Africa and South America largely occupy the role of 'commodity' exporters. UNIDO speaks of a 'clear division' whereby:

59 Luciano Bolinaga and Ariel Slipak, 'El Consenso de Beijing y la reprimarización productiva de América Latina: El caso argentino', *Revista Problemas del Desarrollo* 183(46), 2015, 33–58.

> Industrialized economies, with state-of-the-art technology in their productive units, have given away their labour-intensive operations to other countries while keeping the more knowledge-intensive ones. The industrialized economies are already highly productive and are the fastest to adopt the technology they produce, pushing the technological frontier even further and removing themselves far from the rest of the world. The developing and emerging industrial economies, with most productive units presenting different degrees of technological backwardness, still use low wages as an advantageous entry point for their integration into the global markets.[60]

This account is, however, complicated by the fact that China is developing a strong capacity for endogenous technological development. As the current 'trade war' demonstrates, China is increasingly threatening the Global North's dominance of the technological frontier, having already joined the club of its 'leaders':

> Today's technological breakthroughs in ADP [advanced digital production] are again dividing the world between leaders, followers and laggards. One striking feature of the creation and diffusion of ADP technologies is the extreme concentration, especially of patenting and exporting activity . . . Only 10 economies show above-average market shares in the global patenting of ADP technologies. Ordered by their shares, these economies are the United States, Japan, Germany, China, Taiwan Province of China, France, Switzerland, the United Kingdom, the Republic of Korea and the Netherlands. Together, they account for 91 percent of all global patent families.[61]

Capital and class compositions cannot be separated because they constantly transform each other. New technologies require new skills, and those needed by Industry 4.0 consist of 'analytical skills; specific technology-related skills, including science, technology, engineering

60 UNIDO, *Industrial Development Report 2020*, 151.

61 Ibid., 5.

and math (STEM) – and ICT-related skills; and soft skills . . . The shares of STEM employees are consistently higher among more technologically dynamic firms.'[62]

Marx theorised that automation is based on capital's absorption of both science and workers' skills. As he wrote in the *Grundrisse*: 'In fixed capital . . . skill [is] transposed from direct labour into the machine, into the dead productive force.'[63] Think of the way contemporary artificial intelligence learns from human behaviour and objectifies it into microchips.[64] Workers' 'clandestine' knowledge (following Romano Alquati's terminology), prior to its being incorporated into machines, can only be gained through learning-by-doing on the job.[65] Harry Braverman was correct in claiming that *these* types of informally transmitted skills were made increasingly redundant by capitalist automation.[66] What Braverman did not evidence, however, is that capitalist innovation nonetheless gradually raised the skill threshold of the core workforce, because codified knowledge is required to operate highly automated industrial systems. This 'overt' knowledge needs to be acquired through formal education. There is thus a generalised shift from 'particularistic' to 'universalistic' skills in the core workforce of much large-scale industry.[67] This shift comes at the expense of those workers with little formal education within nations, as well as affecting whole economies with low educational levels internationally.[68]

62 UNIDO, *Industrial Development Report 2020*, 14.

63 Karl Marx, *Grundrisse: Foundations of the Critique of Political Economy*, London: Penguin, 1973 [1939], 715.

64 Nick Dyer-Witheford, Atle Mikkola Kjosen and James Steinhoff, *Inhuman Power: Artificial Intelligence and the Future of Capitalism*, London: Pluto Press, 2019; Matteo Pasquinelli, *The Eye of the Master: A Social History of Artificial Intelligence*, London: Verso, 2023.

65 Romano Alquati, 'Composizione del capitale e forza lavoro alla Olivetti', *Quaderni rossi* 3, 1963, 119–85.

66 Braverman, *Labor and Monopoly Capital*.

67 Greig Charnock and Guido Starosta, *The New International Division of Labour: Global Transformation and Uneven Development*, London: Palgrave Macmillan, 2016, 7–8.

68 Dani Rodrik, 'New Technologies, Global Value Chains, and Developing Economies', National Bureau of Economic Research Working Papers 25164, 2018.

The global tendencies just outlined have deep impacts at a local level, as the life patterns of people, households and whole communities are continually disrupted by the imperatives of capital accumulation. In numerous cases, these broader trends are incarnated through a diverse range of local experiences of noxious deindustrialisation. In such instances, significantly noxious factories continue to operate, but employment in the fenceline communities around them becomes increasingly deindustrialised. First, automation reduces the number of jobs available. Then, because jobs are awarded based on rankings in formal educational proficiency, rather than geographical convenience or informal, localised mechanisms of skills transfer, fewer fenceline community members are able to access the remaining core jobs. Finally, the remaining less skilled jobs are externalised to a partially itinerant and more precarious workforce employed by outsourced firms or temp-work agencies.

Since only higher-income households can easily relocate to healthier areas, fenceline communities around large-scale noxious factories remain disproportionately working class even as they become less industrially employed. They are thus still exposed to the burdens of the polluting industry while receiving less benefits, namely secure jobs, in return. Alongside this, the workforce of large-scale industry becomes less embedded in these communities due to long-range recruiting from further locations. The result is a relative 'deterritorialisation' of the workforce of large-scale industry. Similar trends are even more pronounced around capital-intensive extraction, with 'drive-in/drive-out' or even 'fly-in/fly-out' commuting replacing the company town in high-tech mining sites.[69] As we will see in Chapter 4, the overall result is a *widening of the bifurcation* between workplaces and communities, *between technical composition and social composition.*

69 See Touraj Atabaki, Elisabetta Bini and Kaveh Ehsani (eds), *Working for Oil: Comparative Social Histories of Labor in the Global Oil Industry*, London: Palgrave Macmillan, 2018; Dusan Paredes, Juan Soto and David A. Fleming, 'Wage Compensation for Fly-In/Fly-Out and Drive-In/Drive-Out Commuters', *Papers in Regional Science* 97(4), 2017, 1337–53.

Grangemouth, UK: Glowing fires, vanishing jobs

In July 2023, billionaire businessman Jim Ratcliffe, the CEO and main owner of petrochemical company INEOS, described the UK's energy policy as 'crap', calling for the government to abandon its ban on fracking and to incentivise the further exploitation of North Sea oil and gas.[70] One of the richest men in the UK, the self-styled 'alchemist' never shied away from controversy. His vocal support for Brexit did not stop him from moving his tax residence to Monaco, nor did his much-vaunted working-class background prevent him from crushing the unions in a 2013 dispute in Grangemouth. Ratcliffe's enthusiasm for fossil fuels also brought to Grangemouth protesters from Extinction Rebellion and Climate Camp Scotland, who staged acts of civil disobedience in and around the INEOS premises.

Yet the alchemist's positions are understandable, given that he owns, in addition to many industrial sites, oil and gas licences in the North Sea and beyond. Luckily for Ratcliffe, only a few weeks after his 'crap' declaration, British PM Rishi Sunak announced the government's plans to 'max out' oil and gas drilling in the North Sea. In Ratcliffe's vision, the continued extraction of fossil fuels is necessary to 'put the factories back' in the UK, providing good industrial jobs to British working-class communities (his unilateral closure of Grangemouth's refinery is highly ironic in this respect).[71] However, the paradox of noxious deindustrialisation ensures that such promises are nothing more than a mirage, as exemplified by the fact that the quality of life in Grangemouth was perceived to have declined long before the closure of the refinery. Job losses had, in fact, long been a reality in Grangemouth, even in times of expanding production.

During the post–Second World War boom, the Grangemouth branches of British Petroleum (BP) and Imperial Chemical Industries

70 Rachel Millard, 'INEOS Owner Jim Ratcliffe Calls UK Energy Policy "Crap"', *Financial Times*, 5 July 2023.

71 Jim Ratcliffe, 'I Grew Up amid Manchester's Smokestacks: I Want to Put the Factories Back', 'Rich List' interview by John Arlidge, *Sunday Times Magazine*, 7 May 2017.

(ICI) substantially contributed to local employment and tax revenues, with the town thriving as a result. This constituted the basis of an implicit Fordist social contract between the community and the petrochemical industry.[72] Fordist compromises of this kind were mostly restricted to the Global North and based on manifold exclusions along gendered, racial, imperial and environmental lines (as exemplified by the histories of BP and ICI).[73] In any case, under changing employment and fiscal patterns, this social contract gradually unravelled. In Grangemouth, the very capital of the Scottish petrochemical industry, employment deindustrialisation became a reality despite increasing industrial production. Today, the fenceline community no longer significantly benefits from the industry in terms of either jobs or public services yet is exposed to the socioenvironmental noxiousness of the petrochemical plants that are still operating.

Grangemouth's petrochemical industry dates back over a century. In 1919, Scottish Dyes Ltd built a chemical plant in Grangemouth, which was acquired by ICI in 1928. In 1924, the Anglo-Persian Oil Company, which would become BP thirty years later, established its refinery. The latter's capacity was then augmented in the 1940s and '50s, with further expansion taking place in the 1970s after the discovery of North Sea oil.[74] Grangemouth's petrochemical plants were installed by BP in 1951. The town grew up around its factories, with housing constructed by both the council and BP to shelter a burgeoning workforce.

In the neoliberal period, however, global competition started to exert significant pressure on the relatively favourable conditions enjoyed by Grangemouth's petrochemical workers. In the 1990s, the ex-ICI site was

72 On Fordism, see Brett Neilson and Ned Rossiter, 'Precarity as a Political Concept, or, Fordism as Exception', *Theory, Culture and Society* 25(7–8), 2008, 51–72.

73 Ulrich Brand and Markus Wissen, *The Imperial Mode of Living: Everyday Life and the Ecological Crisis of Capitalism*, London: Verso, 2021 [2017].

74 Terry Brotherstone, 'A Contribution to the Critique of Post-Imperial British History: North Sea Oil, Scottish Nationalism and Thatcherite Neoliberalism', in John-Andrew McNeish and Owen Logan (eds), *Flammable Societies: Studies on the Socio-Economics of Oil and Gas*, London: Pluto Press, 2012, 70–97.

restructured, downsized and later broken up across multiple companies. In 2005, BP ceded its refinery and petrochemical plants to INEOS, a private company founded in 1998 by Jim Ratcliffe, who swiftly turned it into one of the world's leading chemical multinationals. Since 2011, the refinery had belonged to Petroineos, a 50:50 joint venture between INEOS and PetroChina. INEOS also owns the two gas-fired power stations that service the industrial complex.

The overriding theme of the fieldwork I undertook in Grangemouth was the town's decline relative to the three decades that followed the Second World War, when it was known as 'Scotland's Boomtown' due to its rapid economic growth, abundance of good-quality jobs and strong public services. The boom was followed by a multifaceted decline in prosperity, population, community cohesion, reputation, industrial jobs for the locals, employment terms and conditions, tax revenues, public services and transport, housing and infrastructure, industry contributions to the community, and activities in the town centre, to name a just a few facets. This ruination story is strikingly similar to those of former industrial towns all across the world; and yet, industry in Grangemouth was alive and kicking.

The most recurrent grievance concerned the decline in local employment within the petrochemical industry. The evolution of capital composition diminished the number of factory jobs available in Grangemouth, while employment similarly shrunk at the Grangemouth docks due to containerisation. Since its beginnings, the petrochemical sector has been a typical capital-intensive branch of production.[75] However, with the deployment of programmable logic controllers and distributed control systems in the 1970s, followed by advanced digital technologies, the organic composition of capital in the petrochemical sector has increased further.[76] Of the surviving jobs in Grangemouth's petrochemical

75 Adam Hanieh, 'Petrochemical Empire: The Geo-Politics of Fossil-Fuelled Production', *New Left Review* 130, 2021, 25–51; Serge Mallet, *La nouvelle classe ouvrière*, Paris: Editions du Seuil, 1963.

76 Dale E. Seborg, 'Automation and Control of Chemical and Petrochemical Plants', in Heinz Unbehauen (ed.), *Control Systems, Robotics and Automation: Industrial Applications of Control Systems II*, Oxford: EOLSS Publishers, 2009, 57–72.

industry, the less qualified ones have been outsourced to a partially itinerant workforce of contract and agency workers, while the in-house positions now require qualifications that are harder for local residents to achieve. As a retired petrochemical worker told me: 'The qualifications you've got to have to get into that place now are beyond a lot of people.'

Boomtown Grangemouth's slow-motion downfall began with BP's piecemeal privatisation, between 1977 and 1987, and was accompanied by a process of automation and restructuring. The number of direct employees thus shrunk from 5,500 in the 1980s to around 1,300 in 2020.[77] Of these, only a small number resided in Grangemouth: less than 100 in a population of over 16,000, according to an estimate of the Community Council based on data provided by INEOS. A dramatic acceleration in this tendency took place in 2002, when BP abruptly cut its workforce by 700. According to Scotland's 1961 Census data, the share of industrial employment in Grangemouth stood at 55 per cent, with 'chemicals and allied industries' constituting an astonishing 44.5 per cent of total employment. Half a century later, the share of industrial employment had been dashed to 13 per cent, a fraction of what it used to be. This meant a clear change in the local working-class composition, with a decline of secure, industrial employment and an increase in more precarious employment, particularly in the service sector. A petrochemical contract worker commented: 'I'm not saying anyone, but the majority of people in the town don't work in the major industries, the money industries we'll call it. They may work in ASDA [a supermarket] and the non-skilled places.'

Wages and conditions have also generally declined for INEOS's in-house workers, particularly after Unite the Union was defeated in a dispute that culminated in October 2013.[78] INEOS threatened to shut down the complex if its employees did not 'agree' to the withdrawal of their final salary pension scheme, a three-year freeze on wages and

77 Peter Phillimore, Achim Schlüter, Tanja Pless-Mulloli and Patricia Bell, 'Residents, Regulators, and Risk in Two Industrial Towns', *EPC: Government and Policy* 25(1), 2007, 73–89, 76.

78 Mark Lyon, *The Battle of Grangemouth: A Worker's Story*, London: Lawrence and Wishart, 2017.

industrial action, and other degradations in terms and conditions. Eventually, workers were made to reapply for their own jobs on the new terms and the two leading Unite shop stewards, Stevie Deans and Mark Lyon, were respectively pressured to quit and sacked. It was a setback that became emblematic of the relative weakening of a formerly mighty union power.

Disaffection with INEOS was such that an estimated 30 to 40 per cent of direct employees quit following the 2013 dispute, choosing to try their luck in the North Sea and Arabian Gulf. However, this did not result in a significant rise in employment opportunities for the locals, due to the qualifications and experience required in new job postings. Instead, the remaining employees worked longer overtime, while new recruits came from as far as Wales, where the Milford Haven Refinery had recently closed. As shift workers take twelve-hour shifts (day and night), overtime shifts result in extremely protracted working weeks. Moreover, due to the pension scheme retrenchment, early retirement is now a costly option. The overall outcome is a paradox from the workers' perspective, with older employees working long hours and travelling from further afield, while local youths struggle to find employment in the factories on their doorstep.

The downward trend in employment was paralleled by falling tax revenues for Grangemouth. Having been known as 'the Kuwait of Scotland' due to its hydrocarbon-generated wealth, today the presence of the petrochemical industry does not translate into Grangemouth reaping significant extra revenue, as non-domestic rates are pooled by the Scottish government and then redistributed. Additionally, business taxes and taxes on high incomes in the UK have been slashed since the times of the Fordist compromise. Some participants also brought up the subject of tax avoidance: in 2010 INEOS transferred its headquarters from the UK to Switzerland, before partially returning to the UK in 2016 after corporation tax was cut.[79] The company also kept substantial holdings in tax havens such as the Isle of Man, while Jim Ratcliffe himself

79 Helen Miller, *What's Been Happening to Corporation Tax?*, London: IFS, 2017.

moved his residence to tax-free Monaco.[80] The fall in tax income led to a decline in public services, poorer maintenance of the existing infrastructure and a dearth of resources for revitalising the town centre. A community councillor observed that 'there are empty flats that are falling to bits, empty shops and one in four retail properties are empty'.

According to participants, when the jobs and taxes from industry were helping Grangemouth thrive, most locals were happy to co-exist with their difficult petrochemical neighbour. The fenceline community thus learned to live with the permanent hum and night-time orange glow exuding from the complex. However, when the perks began to wane so did patience (even if the environmental record of the industry is accepted to have generally improved over recent decades). The burdens outlined by the participants chiefly related to pollution (including lights, noise, smells, vibrations, smokes, powders and leaks), industrial hazards, social stigma, haulage traffic and blockages to economic diversification and the construction of new housing.

The Scottish Environment Protection Agency (SEPA) revealed that, out of Scotland's top-twelve CO_2 emitting plants, five were located in Grangemouth and owned by INEOS.[81] As a report to the Scottish Parliament remarked: 'The largest point source of emissions in Scotland is now the cluster of industry at Grangemouth, which accounted for over 30 per cent of industry emissions in Scotland (3.6 $MtCO_2$) in 2017.'[82] Among the Grangemouth plants, the Petroineos refinery was responsible for the vast majority of CO_2 emissions, which had slowly been rising since 2014. Air quality was also a crucial concern, as Grangemouth was declared an 'Air Quality Management Area' in 2005 due to sulphur dioxide emissions from the refinery. Another considerable source of complaints was heavy flaring. In sum, the petrochemical industry in Grangemouth remained a major polluter in terms of both greenhouse gases and toxic emissions.

80 'INEOS/UK Tax: Jim'll Switch It', *Financial Times*, 18 February 2019.

81 Rob Edwards, 'Named: The "Dirty Dozen" Companies That Pollute Scotland', *Ferret*, 6 January 2019, theferret.scot.

82 Committee on Climate Change, *Reducing Emissions in Scotland: 2019 Progress Report to Parliament*, London: Committee on Climate Change, 2019, 63.

In addition to industrial hazards, several participants worried about suspected health effects – particularly cancers and respiratory diseases – and expressed frustration over the hopelessness of verifying them. In fact, there were no available epidemiological studies on the health impacts of industry for Grangemouth specifically. Furthermore, noxious deindustrialisation appeared to affect mental health through precarious employment patterns and weakening community ties. For instance, a participant in his thirties claimed that 'the mental health in the area has plummeted. I mean as folk were saying about the social clubs, now even the pubs are dead, so nobody is being brought together for anything.'

While most interviewees would probably not have described themselves as 'environmentalists', some added concerns about the role of the industrial complex in global environmental degradation, citing climate change and plastic waste pollution. An example of converging local and global environmental demands related to fracking. INEOS consistently lobbied public authorities to deregulate fracking, and it was the largest holder of UK fracking licences. However, due to the Scottish government's moratorium, INEOS imported fracked ethane from Pennsylvania, US by ship. Several participants lambasted the prospect of the additional environmental damage and hazards fracking would engender: 'The only customer for the fracked gas is the INEOS plant. That's it. They want to destroy our countryside, put old people at their knees, your insurance goes through the roof and all these earthquakes that are happening, like in Blackpool, and deliveries coming in and fire and all the rest of it, for one man's greed.'

A range of economic burdens were also mentioned. In fact, current regulations set limits on the siting of new economic activities in the 'consultation zones' (also known as 'blast zones') within a certain vicinity of hazardous industries. Furthermore, a sizeable portion of Grangemouth's housing stock is constituted by flats that were built as council houses. According to participants, these accommodations no longer appeal to the core petrochemical workforce. Those flats that are still council-owned are assigned to marginalised people from elsewhere in Scotland who, due to budget constraints, are not given adequate support to integrate. This leads to perceived high incidences of drug abuse and anti-social behaviour and contributes to the higher-than-average presence of surplus workers – a common characteristic of communities located close to large-scale noxious factories.

Interviewees knew that INEOS was not the only polluting industry on their fenceline, but they tended to single it out as a target of discontent. This was confirmed in an independent study commissioned by Falkirk Council: 'Of all industry that operates within Grangemouth, INEOS was overwhelmingly seen as negative.'[83] In Grangemouth, post-war nostalgia thus overlaps with BP nostalgia, as signalled in the expressions used by focus group participants in 2019: during 'the golden years', BP was 'something to be proud of', 'a good company to work for', 'helpful and supportive', 'family oriented', 'part of the community', 'at a different level', 'just incredible', and 'you know, great'. Comparison with the descriptors reserved for INEOS is telling: 'very bad neighbours', 'secretive', 'hard-headed' and 'ridiculous'. Jim Ratcliffe meanwhile was described as 'a loose cannon', 'very immoral', 'a bully-boy' and 'an absolute rat'.

The contrast between Ratcliffe's confrontational attitude and extravagant wealth versus Grangemouth's demoralisation and 'abject poverty' (in an interviewee's words) engendered strong feelings of injustice. While Ratcliffe was declared the UK's richest individual in 2018 with a personal fortune of over £21 billion, Grangemouth includes five areas that fall within the most deprived 10 per cent of Scotland.[84] Widespread fuel poverty was described as particularly paradoxical given the community's proximity to the refinery: 'We're giving food parcels out and people say, "Is there anything we can eat cold?" because they can't afford the power to heat it up, so what do you do? I mean it is heart breaking.'

A widespread view was that Ratcliffe possessed both disproportionate wealth and power. In little more than a decade, INEOS obtained the infrastructure that produced 4 per cent of Scotland's GDP, refined 80 per cent of its fuel and delivered almost 40 per cent of North Sea oil and

83 Community Links, 'Grangemouth "Your Place Your Views" Consultation: Final Report', 2019, 27.

84 Falkirk Council, 'Scottish Index of Multiple Deprivation 2020', falkirk.gov.uk, 2020.

gas.[85] Furthermore, INEOS is a privately owned company and thus its majority owner and CEO, Ratcliffe, is not bound by the same accountability requirements as public companies (not to mention state-owned ones). Frustration at this concentration of power often emerged in our focus groups: 'It would bring tears to your eyes to hear what this town is like for the people who live in it, and I guess again, some of that is coming back to INEOS holding the town to ransom'; 'They've also got the North Sea pipeline now, so he's got a stranglehold on Scotland really'; 'Why the Scottish government allowed that to happen, I don't know.' The latter comment indicated a scepticism towards public authorities – seeing them as unwilling or unable to confront industry. This was displayed by several respondents, including one public official: 'They don't have the authority, so it's like this big beast is just out of control.'

A petrochemical contract worker summarised the experience of noxious deindustrialisation as follows: 'I've lived in Grangemouth for all my days . . . I watched the industry go downhill and what I feel more about INEOS now is there's more people coming into the town to work and the people in the town can't get the jobs . . . We've got to put up with the noise, the smell and the danger most of all. I mean I'm two minutes from the refinery and the noises and the sky lighting up at night can be quite scary . . . I shouldn't have to travel away from the town to get work when that's on my doorstep. I just don't think that it's right.'

The interacting technological and socioeconomic changes just described have seriously strained community–industry relations: 'It was a boomtown. Well, no longer it's a boomtown . . . Technology changed and now there are so many programmable logic controllers and then, of course, they outsourced a lot of their staff to contractors. Now, they used to have boilermakers, electricians, fitters and everything in-house'; 'What we have now is what we refer to as DIDOs or Drive In and Drive Out. People drive into work and they drive out at night. They take their well-earned cash with them and they spend it elsewhere.' In sum, the picture drawn by interviewees was one of frustration and decline, with

85 Natalie Thomas, 'INEOS Buys North Sea Oil Pipeline from BP', *Financial Times*, 3 April 2017.

the fear that the area was turning from 'Scotland's Boomtown' into 'Scotland's dumping ground' repeatedly surfacing.

This fraying of the Fordist community–industry social contract is not a phenomenon exceptional to Grangemouth but can be observed in many other cases. On the one hand, the co-evolution of capital and class compositions has tended to result in a decline in the quantity and quality of industrial jobs available to locals and a related increase in precarious employment. On the other hand, the ongoing environmental degradation and a heightened awareness of health and environmental issues have increased the perceived costs of industry.

Local concerns about large-scale noxious sites are often dismissed as selfish and in contradiction with the general interest, as the acronym Not in My Backyard (NIMBY) indicates. However, the dystopian picture painted by noxious deindustrialisation, with its corollaries of steep inequality, employment precarity, cumulative environmental degradation and weakening community ties, is something one can be forgiven for being worried about. And even more so when we recognise the very real global possibility that a combination of automation and slow economic growth will continue to exacerbate noxious deindustrialisation. While many long for a return to a romanticised golden age of industrial capitalism, this is both impossible and undesirable. In fact, the output growth necessary to reindustrialise employment is now incompatible with the sustained reproduction of life on the planet. To conclude with the words of a participant: 'Grangemouth has already transitioned away from the high-paid jobs, so we've started that transition and it's not been just . . . We need something else.'

2

The Surplus Working Class: Precarity in Environmental Degradation

In a cartoon from her satirical series *Willis from Tunis*, comic artist Nadia Khiari depicted the then president of Tunisia Beji Caid Essebsi being asked about solutions to youth unemployment, which was standing at 35 per cent. Essebsi, portrayed with a catlike appearance, replies: 'They should grow old, quickly.' However, the social duration of youth is not only defined by age, as its end incorporates features such as bearing family responsibilities, having independent accommodation and, above all, being in a steady job.[1] This norm was aptly conveyed by Roman cartoonist Zerocalcare, who wrote:

> 'Thirty-year-old', in Italian, does not refer merely to a cold registry record. 'Thirty-year-old' applies to a family of large mammals bearing traits of maturity, emancipation, stability. This species became extinct decades ago, more or less since Treu's labour market reform.[2]

1 See Hadas Weiss, 'Elusive Adulthood and Surplus Life-Time in Spain', *Critique of Anthropology* 41(2), 2021, 149–64.

2 Zerocalcare, *La profezia dell'armadillo*, Milan: Bao, 2011. Tiziano Treu is a centre-left Italian politician. In 1997, as Minister of Labour in Romano Prodi's government, he promoted a set of laws liberalising the labour market.

A consequence of employment precarity, defined here as dependence on and irregularity in accessing a monetised income, is the lengthening of the average social duration of 'youth', understood as the transition phase from childhood to adulthood. As shown, for example, by Stefano Pontiggia's ethnography of marginality in Tunisia's Gafsa mining basin, this delay of adulthood is often experienced as an insufferable curse.[3] Such an anxiety is so widespread that the emblematic track 'Houmani' by Tunisian rappers Hamzaoui Med Amine and Kafon – lamenting that 'by their nature, pennies don't stay long in my pocket' – quickly rocketed beyond 13 million YouTube clicks, more than Tunisia's entire population.

But youth is also the time when experiences are most valuable and intense, relationships create the strongest bonds, and the future is unwritten. The problem, then, is not the extension of youth itself. An idealised Fordist model, which never applied to more than a mostly white and male minority of the global population, set the standard for 'proper' working-class adulthood around the world.[4] However, even Fordist workers did not enter the factories in their teens because they wished to spend their best years on an assembly line. Rather, as a recent viral meme sharply pointed out, they did so because they were 'passionate about not starving to death' and having pennies stay in their pockets for a bit longer.

What turns the postponement of adulthood into an insufferable curse for large chunks of the global working class is the lack of personal and collective autonomy in which youth is exercised. Precarity is not a choice, it is a condition externally imposed by the impersonal forces of the labour market. This often happens in the absence of the economic and legal means to look elsewhere for new experiences, so that they can be sought only at the cost of risking one's life in dangerous journeys as an illegal migrant. If many of the workers who crossed the factory gates or

3 Stefano Pontiggia, *Revolutionary Tunisia: Inequality, Marginality, and Power*, Lanham, MD: Lexington, 2021.

4 Sarah Mosoetsa, Joel Stillerman and Charles Tilly, 'Precarious Labor, South and North: An Introduction', *International Labor and Working-Class History* 89, 2016, 5–19.

stayed home doing housework since their teens feel like they are spending their best years within prison walls, then many 'workless' workers sense that they are spending their best years in wall-less cages. And climate change, engendered by the very market forces that built these open-air prisons for the surplus working class, is quickly turning up the heat on them.

In English-speaking academia, the best-known deployment of the precarity concept is Guy Standing's *The Precariat.*[5] Standing frames those in precarious employment as a separate class, outside the working class. However, the term working class itself can be defined in manifold ways and definitions cannot be proved or disproved by empirical evidence, such that different conceptions of class can only be judged by their political effectiveness. A useful definition of working class needs to point to a set of potentially common interests that can push forward social change. To complicate things further, the statement that something is in someone's interest, even if it does not entirely coincide with their actually expressed preferences, is itself not an empirical description but a political proposition. Yet, there must be a relationship between preferences and interests for claims about the latter to be credible.

On these grounds, and *contra* Standing, Erik Olin Wright made a solid case for seeing precarious and secure workers as part of the same class because of their overall convergence of interests versus the capitalist class.[6] In fact, the dispossession shared by all workers points to a potential interest in democratising ownership and control of the means of production. This does not then mean that all workers' interests can be reduced to their dispossession, as there is a wide range of interests that cuts across the working class, based for instance on gendered and racialised hierarchies. Instead, it means that even if the dispossessed are divided in many respects, they could find common ground in the

5 Guy Standing, *The Precariat: The New Dangerous Class*, London: Bloomsbury, 2011.

6 Erik O. Wright, 'Is the Precariat a Class?', *Global Labour Journal* 7(2), 2016, 123–35.

overcoming of their dispossession, with the empowerment of the workers at the bottom of the gender-race-class hierarchy a crucial condition for advances in this direction.[7]

For Marx, as already noted, automation augments the productivity of the workers who remain in core production, tending to make an increasing proportion of the working class *relatively* less necessary to accumulation.[8] Again, the distinction between 'relative' and 'absolute' is key here, as capitalism still absolutely depends on precarious workers, precisely because their insecurity contributes to its stabilisation. This surplus working class puts latent or active pressure on the relatively secure workers, eroding their bargaining power and thus blurring the sociological frontier between the precarious and the secure workforce. In our world of employment deindustrialisation, precarity spreads across sectors and subjects, undermining or weakening the forms of consciousness and organisation that characterised the age of expanding industrial employment rates. Of course, rising precarity is not a deterministic 'law' any more than the rising organic composition of capital is. Indeed, after the Second World War many countries saw the advent of a social pact providing sources of regular income for some sections of the population. This period, however, can be regarded as an exception in the history of capitalism and one dependent on a series of specific conditions. Since the 1970s, the social pact has continuously deteriorated, and no inversion of this tendency is in sight.

Precarity and ecological crisis are mutually reinforcing. On the one hand, the scarcity of good jobs encourages a craving for capitalist growth, opening the way for new rounds of environmental damage, austerity and restructurings. On the other hand, the ecological crisis acts as a vector of further dispossession, turning more people into precarious workers, while making precarity itself more unbearable. In fact, the more workers are precarious, the more they tend to be exposed to the impacts of

7 Gargi Bhattacharyya, *Rethinking Racial Capitalism: Questions of Reproduction and Survival*, Lanham, MD: Rowman & Littlefield, 2018.

8 Karl Marx, *Capital: A Critique of Political Economy*, vol. 1, London: Penguin, 1976 [1867], 762–860.

environmental degradation. As Anna Tsing wrote: 'One half of current precarity is the fate of the earth . . . the other half . . . [is] that everyone depends on capitalism but almost no one has what we used to call a "regular job."'[9]

The pincer movement

Considering precarious workers as part of the total working-class composition is not particularly heterodox. Marx himself shifted from seeing the 'lumpenproletariat' as a separate, 'dangerous class' in the *Communist Manifesto* to analysing the surplus population as a segment of the working class in *Capital*. I thus use a definition of working class based on dispossession rather than exploitation. Simply put, unemployed people are not exploited, but most of them are still part of the working class. Dispossession is defined as the lack of ownership and control of *significant* magnitudes of means of production. To live, the dispossessed must acquire money to buy means of subsistence on commodity markets, which is usually done by working. This definition of working class then includes all those who face the compulsion to sell their labour power to produce commodified goods or services, or reproduce other labour power, no matter how successful they are in finding a buyer.[10]

However, not all labour-power sellers are limited to producing commodities or reproducing labour power. In fact, the capitalist class delegates some control responsibilities to the middle class, including the oversight of people who are excluded from supervisory power (and usually remunerates this accordingly). The middle class is thus made of people who perform significant shares of both worker functions (production or reproduction) and capitalist functions (ownership or

9 Anna L. Tsing, *The Mushroom at the End of the World: On the Possibility of Life in Capitalist Ruins*, Princeton, NJ: Princeton University Press, 2015, 3.

10 See Michael Heinrich, *An Introduction to the Three Volumes of Marx's* Capital, New York: Monthly Review Press, 2004, 44.

control). In this sense, middle-class people are simultaneously both workers and capitalists in varying proportions.[11]

While it excludes the middle class, the conception of working class deployed here is still broader than dominant understandings – widened to include the unemployed, reproductive workers, informal workers, subordinated cognitive workers and dependent self-employed workers.

The global working-class composition thus understood is marked by gendered and racialised segmentations, and its surplus component is a concentration of these stratifications.[12] *Operaismo*, for all its strengths, did not have much to say on this. In fact, its most well-known authors focused on the working-class segments they deemed to have the mightiest disruptive power, as exemplified by Tronti's famous quote: 'Today the chain must be broken not where capital is weakest, but where the working class is strongest.'[13] In its most simplified version, this schema paired each phase of capitalist accumulation with a hegemonic 'revolutionary subject', deemed to be the 'strongest' working-class segment.[14] First came the 'professional worker', then the 'mass worker', followed by the 'social' or 'cognitive worker' who closed the trinity.[15]

The *professional worker* of the nineteenth century still mastered the labour process, was proud of his (the male adjective is not casual) skills and knowledge, and thus aspired to the self-management of production.[16] The revolutionary cycle inaugurated in 1917, spearheaded by

11 Erik O. Wright, *Class Counts: Comparative Studies in Class Analysis*, Cambridge: Cambridge University Press, 1997, 20.

12 Sean Hill II, 'Precarity in the Era of #BlackLivesMatter', *Women's Studies Quarterly* 45(3–4), 2017, 94–109; Helena Hirata, 'A precarização e a divisão internacional e sexual do trabalho', *Sociologias* 21, 2009, 24–41.

13 Mario Tronti, *Workers and Capital*, London: Verso, 2019 [1966], 39.

14 For a critique of this schema from the perspective of class composition analysis, see Salvatore Cominu, 'Che cos'è la composizione di classe?', *Machina Rivista*, 5 April 2023, machina-deriveapprodi.com.

15 Antonio Negri, 'Proletarians and the State: Toward a Discussion of Workers' Autonomy and the Historic Compromise' [1975], in *Books for Burning: Between Civil War and Democracy in 1970s Italy*, ed. Timothy S. Murphy, London: Verso, 2005, 118–79.

16 Sergio Bologna, 'Class Composition and the Theory of the Party at the Origin of the Workers-Council Movement', *Telos* 13, 1972, 4–27.

workers' councils taking over the factories, was the political apex of this class composition. This movement, however, was eventually defeated by the 'technological path to repression', powered by electrification (e.g. the assembly line) and eased by state planning geared towards saving capitalism from its own contradictions.[17]

As Marx had already observed from the most advanced developments of his time, the absorption of human skills into fixed capital turned the latter into an 'animated monster', with the worker reduced to an 'individual punctuation mark . . . its living isolated accessory'.[18] This *mass worker*, described by Tronti as a 'rude pagan race, with no ideals, faith, or moral', was typically a recently proletarianised peasant-in-transition, thrown into factories designed to be manned by unskilled labourers. Relative to the professional worker, the mass worker had a more naked experience of how the capitalist labour process tends to empty work of intrinsic meaning and was thus more prone to the 'refusal of work'. Such a refusal culminated in the 'Long 1968' cycle of struggles. Once again, defeat came at the hands of technological transformation coupled with a redeployment of the state to absorb the contradictions of capitalist accumulation. This time, it happened through both the diffusion of digital technologies, which afforded the fragmentation of production and the scattering of value chains all over the world, as well as the neoliberal retreat of the state from direct planning, which depoliticised the sacrifices demanded of the working class.

An example of the difference between the professional worker and the mass worker can be found in Veneto, the Italian region that incubated an academic nucleus of *operaistas* at the University of Padua. There, the professional worker was represented by Venice's skilled craftsmen who, even when forced to lend their dexterity to capitalist factories, carried

17 Antonio Negri, 'Keynes and the Capitalist Theory of the State Post-1929' [1968], in *Revolution Retrieved: Writings on Marx, Keynes, Capitalist Crisis and New Social Subjects (1967–83)*, ed. John Merrington, London: Red Notes, 1988, 5–42, 6. This is the last article by Negri mainly informed by *Capital*, before his turn to the 'Fragment on Machines'.

18 Karl Marx, *Grundrisse: Foundations of the Critique of Political Economy*, London: Penguin, 1973 [1939], 470.

with them a tradition dating back to medieval times, centred around the Arsenal shipyard and Murano's glassworks. In 1920, the Arsenal workers joined the movement of factory takeovers, and during the Second World War they supported the anti-fascist Resistance in significant numbers.[19] Negri recounted that one of Porto Marghera's first *operaista* factory militants was Vetrocoke's Giuseppe Pistolato: 'Vetrocoke was the factory where, alongside the coke-based chemical industry, the best crystals in Europe were made. It was rumoured that its skilled workforce had been trained in Murano's glassworks.'[20] In 1963, faced with a managerial attempt to technologically upgrade the labour process, the workers 'destroyed the factory, they blew up the glass furnaces not to lose the dignity of their craft'.[21] It was the Venetian professional worker's last stand.

The Veneto mass worker, in contrast, was incarnated by the less 'class conscious' peasants-in-transition, coming from the hinterland to crowd Porto Marghera's dirtiest, semi-automated chemical factories.[22] Some of them, the so-called '*metalmezzadri*' (literally 'metal-sharecroppers'), still had access to small plots of land. Until the 1950s, their condition was indeed extremely precarious, albeit mitigated by varying degrees of lingering access to the land.[23] However, in the 1960s, for a fleeting but defining moment, these 'peasant-factory workers' became the main social basis for *operaista* intervention.[24] Job security only arrived as a consequence of these struggles, with the newly conquered guarantees codified in the Workers' Statute of 1970. By then, however, neoliberal restructuring was already beginning to surface.

19 Cesco Chinello, 'Storia operaia di Porto Marghera', Treccani Enciclopedia, 2002, treccani.it; Isabella Peretti, *Lotte operaie a Porto Marghera durante la Resistenza*, Venice: Comitato zona industriale PCI, 1972.

20 Antonio Negri, 'Un intellettuale tra gli operai', in Devi Sacchetto and Gianni Sbrogiò (eds), *Quando il potere è operaio: Autonomia e soggettività politica a Porto Marghera, 1960–1980*, Rome: Manifesto Libri, 2009, 140–50, 142.

21 Ibid.

22 Francesco Piva, *Contadini in fabbrica: Il caso Marghera, 1920–1945*, Rome: Edizioni Lavoro, 1991.

23 Francesco Piva and Giuseppe Tattara (eds), *I primi operai di Porto Marghera: Mercato, reclutamento, occupazione, 1917–1940*, Padua: Marsilio, 1983.

24 Sacchetto and Sbrogiò, *Quando il potere è operaio*.

The *post-operaista* theory of the *cognitive worker* can be seen as a wager on applying the 'hegemonic worker' schema to the neoliberal phase of capitalism. In this narrative, the refusal of factory work spurred the further automation of many manual tasks, which increased the need for engineers and computer scientists. At the same time, the demands for free time, and more self-fulfilling uses of it, expanded employment in the creative industries. As a result of these conflictual processes, the cognitive worker was thought to be the figure with the highest potential to lead the struggle in the new times.[25] In this context, *post-operaismo* delivered an understanding of precarity as rooted in the decline of Fordist standardised manufacturing and the rise of 'immaterial' labour.[26] In the 2000s and 2010s, publications such as *Quaderni di San Precario* provided valuable theoretical elaborations and militant inquiries, which assisted Italian social movement organisations and radical unions in organising segments of the working class that were difficult to reach for the traditional labour movement. Gratitude for these contributions is thus owed to them.

However, as argued in the introduction, the thesis of a hegemony of 'immaterial' labour as proposed by Hardt and Negri has its problems. In fact, the cognitive workers failed to live up to the revolutionary burden placed upon their shoulders, and even more spectacularly so than their professional and mass ancestors. Cognitive workers were an important component of the 'alter-globalisation' movement, the post-2008 anti-austerity struggles and the anti-patriarchal, anti-racist and environmentalist mobilisations of the 2010s. However, it takes a heroic simplifying effort to pigeonhole the revolts of underemployed youths, anti-extractivist and peasant struggles (think of the Zapatistas), the strikes of factory, logistics and 'gig economy' workers, and many more under the hegemony of the cognitive worker. While educational levels have risen over time, in many contexts opportunities for qualified employment

25 Michael Hardt and Antonio Negri, *Empire*, Cambridge, MA: Harvard University Press, 2000, 52–3.

26 Michael Hardt and Antonio Negri, *Multitude: War and Democracy in the Age of Empire*, London: Penguin, 2004, 66.

have not increased at the same rate. The weight of 'immaterial' labour is therefore limited and cannot provide the main infrastructure for mobilisation.

Crucially, the thesis of cognitive worker hegemony also fails to appropriately recognise the role of gendered reproductive labour and racialised super-exploited labour, as well as the importance of struggles against them. This provides an occasion to scrutinise the limits of the original *operaista* schema too, at least in its simplified interpretation. In fact, the theories of the professional and mass workers suffered from similar blind spots. The fact that white male theorists tended to identify white male workers as the 'strongest' working-class segments should be looked upon with suspicion.[27] The analyses of the professional, mass and cognitive worker trinity have their validity for these specific working-class segments, along with many brilliant insights. However, what needs to be rejected, today as well as historically, is the hegemonic role assigned to them. The point, then, is not to look for the new 'working-class hero' who will lead the whole proletariat to victory. It is, rather, to strive collaboratively to broaden class composition analysis to all segments of the global working class; looking for ways to weave convergences across positionalities, industries and geographies. Casting aside stagist interpretations of working-class transformation also allows us to see today's professional and mass workers not as mere residues of past eras, but as the living elements of labour processes that are still part and parcel of total capital accumulation.

That said, *operaismo* as a whole was not blind to the struggles of working-class segments beyond the supposed 'hegemonic trinity'. Indeed, the *operaista*-inspired press was full of reports about anti-imperialist struggles in the Third World or anti-racist organising in the US. The critical juncture of the 1970s was seen as resulting from both the insubordination of workers in the core and national liberation struggles in the

27 Silvia Federici, 'Precarious Labour: A Feminist Viewpoint', *In the Middle of the Whirlwind* (blog), 2008, inthemiddleofthewhirlwind.wordpress.com; Ramón Grosfoguel, 'Del imperialismo de Lenin al Imperio de Hardt y Negri: "Fases superiores" del eurocentrismo', *Universitas humanística* 65, 2008, 15–26.

periphery, as 'the working-class refusal of exploitation and the armed resistance of the Vietnamese jointly precipitated the crisis'.[28] Furthermore, writers who have been somewhat marginalised from the *operaista* canon directed their attention beyond the white mass worker. For example, Ferruccio Gambino developed an analysis of migrant and racialised labour inspired by the work of authors such as C. L. R. James and Grace Lee Boggs, while Luciano Ferrari Bravo and Alessandro Serafini explored debates on imperialism to make sense of working-class migration (the 'multinational working class', as they called it) and uneven development within Italy.[29]

Nevertheless, those who best rose to the challenge of criticising mainstream *operaismo*'s blind spots were its feminist renegades.[30] As detailed in Chapter 4, authors such as Mariarosa Dalla Costa, Alisa Del Re and Leopoldina Fortunati showed that unwaged housework, which the gendered division of labour mostly assigns to women, is integrated into capital accumulation because it produces and reproduces labour power. Housework is a special type of precarious work, in the sense that it guarantees an income for only as long as the waged member of the household is willing to be the family's breadwinner. The 'employment relationship' between the house worker and the waged household member who brings home a monetary income is, to a large extent, individualised and unregulated, which increases the exposure of house workers to the risk of gendered violence. Selma James, whose

28 Sergio Bologna, 'Petrolio e mercato mondiale', *Quaderni piacentini* 52, 1974, 3–27, 4.

29 Ferruccio Gambino, *Workers' Struggles and the Development of Ford in Britain*, London: Red Notes, 1976 [1972]; Luciano Ferrari Bravo, 'Old and New Questions in the Theory of Imperialism' [1975], *Viewpoint Magazine*, 1 February 2018, viewpointmag.com; Luciano Ferrari Bravo and Alessandro Serafini, *Stato e sottosviluppo: Il caso del Mezzogiorno Italiano*, Milan: Feltrinelli, 1973.

30 Lucia Chisté, Alisa Del Re and Edvidge Forti, *Oltre il lavoro domestico: Il lavoro delle donne tra produzione e riproduzione*, Milan: Feltrinelli, 1979; Mariarosa Dalla Costa, *Women and the Subversion of the Community: A Mariarosa Dalla Costa Reader*, Oakland, CA: PM Press, 2019; Leopoldina Fortunati, *The Arcana of Reproduction: Housewives, Prostitutes, Workers and Capital*, London: Verso, 2025 [1981].

relationship with this strand of feminism deteriorated over time, also brought race into the analysis.[31]

The insights of this feminist tradition not only shed light on invisibilised female work, but also on the capital–labour relationship in its totality. Building on the work of Global South Marxists such as Samir Amin and Frantz Fanon, the renegades of *operaismo* feminism made crucial contributions to disturbing the spurious equivalence between the working class and the wage.[32] Ultimately, there is a wide array of ways through which capital bypasses the formal wage relation, from commodified slavery and dependent self-employment to unwaged housework.[33] As unwaged and intermittently waged work regimes have been disproportionately applied to racialised and gendered workers, steady wages tend to be associated with masculinity and whiteness.[34]

Precarious work is also unevenly distributed across the planet, as pointed out in the 'marginality debate' of the 1960s and '70s in the context of Latin American dependency theory.[35] The latter highlighted that, as a result of colonisation, the Global North dominates the technological frontier while the Global South is inserted in the world economy in a subordinate position. Global South economies thus tend to be characterised by the deployment of exogenously developed machineries and organisations of work, imported from the Global North after they cease to be cutting edge there. However, as these imported technologies are

31 Selma James, *Sex, Race and Class: The Perspective of Winning. A Selection of Writings (1952–2011)*, Oakland, CA: PM Press, 2012.

32 Samir Amin, *Accumulation on a World Scale: A Critique of the Theory of Underdevelopment*, New York: Monthly Review Press, 1974 [1970]; Frantz Fanon, *The Wretched of the Earth*, New York: Grove Press, 2021 [1961]; Silvia Federici, *Revolution at Point Zero: Housework, Reproduction, and Feminist Struggle*, Oakland, CA: PM Press, 2012, 6–7.

33 Jairus Banaji, 'The Fictions of Free Labour: Contract, Coercion, and So-Called Unfree Labour', *Historical Materialism* 11(3), 2003, 69–95.

34 Silvia Federici, *Patriarchy of the Wage: Notes on Marx, Gender, and Feminism*, Oakland, CA: PM Press, 2021; Aníbal Quijano, 'Coloniality of Power and Eurocentrism in Latin America', *International Sociology* 15(2), 2000, 215–32.

35 See Vania Bambirra, *Teoría de la dependencia: Una anticrítica*, Mexico City: Era, 1978; Andre G. Frank, 'The Development of Underdevelopment', *Monthly Review* 18(4), 1966, 17–31.

relatively capital intensive, they absorb a limited amount of workers and require specific skills, unrelated to the past experiences and knowledge of most local workers. Previous labour processes therefore do not disappear, but are rather modified, downgraded and subordinated to the new ones. As Aníbal Quijano noted:

> The workforce which is formed and displaced from rural sectors, and that which is formed and displaced between urban sectors, finds itself trapped in a veritable pincer movement: a set of primary productive sectors [mainly agriculture] incessantly displace labour-power, which grows at very quick demographic rates, and a set of urban productive sectors, whose higher development levels limit their quantitative demand for labour, increase their qualitative requirements [skill thresholds], concentrate their market in just a few centres, whilst the low and intermediate levels keep losing access to productive resources.[36]

Quijano indicated that his understanding of the 'marginal pole' of the working class was not in opposition to Marx's theory of the surplus population, but complementary to it. However, Ruy Mauro Marini built an understanding of this question more firmly steeped in Marxist categories.[37] He pointed out that, in the core, class struggle could be integrated into the accumulation of relative surplus value.[38] At the time of the Fordist compromise, productivity gains cheapened the wage goods of Global North workers, raising their real wages, expanding consumer markets and propelling further innovation. In the periphery, however, this route was hindered by limited capabilities for endogenous technological development. While the production of relative surplus value is far from impossible in the periphery, absolute surplus value (based on long hours, intense work and limited working-class consumption, and

36 Aníbal Quijano, 'The Marginal Pole of the Economy and the Marginalized Labour Force' [1970], *Economy and Society* 3(4), 1974, 393–428, 414.

37 See Gil Felix, 'On the Concept of the Reserve Army of Labor in Ruy Mauro Marini', *Latin American Perspectives* 49(1), 2022, 75–90.

38 Ruy M. Marini, *The Dialectics of Dependency*, New York: Monthly Review Press, 2022 [1973].

enforced through low job security) takes a larger share of total accumulation in comparison to the core. In this scenario, a significant part of the productivity gains achieved in the Global South are functional to guaranteeing *relatively* high consumption levels to Global North workers, rather than to Global South workers themselves (think of how the Global North imports cheap food from Latin America or low-cost manufactured goods from Asia).

These obstacles to the expansion of internal consumer markets in the periphery explain the fact that, in much of the Global South, jobs 'are structured in an unstable and precarious way' and are unequally allocated along the racial hierarchies drawn by coloniality, the continuing legacy of colonisation.[39] Quijano wrote that 'classes in Latin America have color'.[40] One may add that the global working-class composition has colour, as it is internally segmented along race lines that function as vectors of stratification and division.[41] Furthermore, the end of the Fordist compromise in the capitalist core coincided with an increase of labour migrations from South to North. The consequences of rising employment precarity in the Global North have also had a colour, with diasporic workforces disproportionately allocated to precarious employment.

The 'pincer movement' described by Quijano – in which, on the one hand, peasants, herders and artisanal fishers are forced away from their livelihoods but, on the other hand, cannot find steady employment elsewhere – has an ecological dimension too. In fact, environmental degradation is one of the drivers of the 'new enclosures', while the consequently enlarged surplus working class tends to be over-exposed to its impacts.[42] Think of a peasant whom climate change has contributed to force off the land (the first swing of the pincer movement), who then works under

39 Quijano, 'The Marginal Pole of the Economy', 404; Aníbal Quijano, 'América Latina en la economía mundial', *Ecuador debate* 31, 1994, 87–100.

40 Quijano, 'Coloniality of Power and Eurocentrism', 230.

41 See Cedric Robinson, *Black Marxism: The Making of the Black Radical Tradition*, London: Zed Books, 1983.

42 On the 'new enclosures', see Midnight Notes Collective (eds), *The New Enclosures*, New York: Autonomedia, 1990.

scorching heat in the informal construction sector (the second swing). This often applies, for example, to the migrant workers building the shining cities of the Arabian Gulf petro-monarchies. The ILO notes that 'the most vulnerable workers in developing and emerging countries (e.g. the self-employed in agriculture or migrant workers in the construction sector) are the hardest hit by heat stress'.[43]

David Harvey used the phrase 'accumulation by dispossession' to designate the continuation of primitive accumulation throughout the course of capitalist development.[44] Enduring dispossession is here a condition for further accumulation. However, by deepening the ecological crisis, capital accumulation itself constitutes a driver of dispossession. Paraphrasing Harvey, the ecological enclosures engendered by capitalist noxiousness can be seen as 'dispossession by accumulation'. Accumulation by dispossession and dispossession by accumulation weave a self-reinforcing spiral swelling the ranks of the surplus working class by depriving direct producers of their livelihoods. The expansion of frontiers of extraction, the diversion of scarce hydric resources to monocrop and mining, the contamination of land and water, as well as droughts and extreme weather events all contribute to dispossess peasant, fishing and pastoralist populations. In a context of global employment deindustrialisation, these people do not then necessarily convert into new industrial working classes. They often become semi-proletarianised surplus workers, locked up in a limbo between a degraded subsistence livelihood and precarious work in the money economy.[45] Peasants-in-limbo, rather than peasants-in-transition.

Militarised borders between the Global North and South are creating a racialised 'climate apartheid', in which some sections of the global

43 International Labour Organization, *Working on a Warmer Planet: The Impact of Heat Stress on Labour Productivity and Decent Work*, Geneva: International Labour Office, 2019, 18.

44 David Harvey, 'The "New" Imperialism: Accumulation by Dispossession', *Socialist Register* 40, 2004, 63–87.

45 Sam Moyo, Paris Yeros, and Praveen Jha, 'Imperialism and Primitive Accumulation: Notes on the New Scramble for Africa', *Agrarian South* 1(2), 2012, 181–203.

working class are promised protection from the worst impacts of environmental degradation, while the others are left to fend for themselves.[46] These dynamics have found their most brutal iteration in the racialised apartheid imposed by the Israeli government on the Palestinian people and land.[47] In the West Bank, the Israeli state and the illegal settlements have dispossessed the indigenous from the most coveted hydric and agricultural resources, often using environmental conservation as a tool for legitimisation.[48] The workers severed from the land have then been partially inserted in the Israeli labour market, for example in construction, in a subordinate position relative to the Israeli working class. Meanwhile, the Gaza Strip has been turned into an open-air prison for a mostly surplus working class, dependent on rents from the outside for its subsistence. At the time of writing, tens of thousands of Palestinians have been exterminated by Israeli carpet bombing, which will also have long-term consequences for the ecological conditions of reproduction on that soil.

With high precarity comes high exposure to noxiousness. Fear of losing one's income pressures precarious workers to accept more dangerous and demanding tasks, for which they often receive little training, alongside the greater stress that the vagaries of the labour market puts on their mental–physical health continuum.[49] Precarious workers also tend to lack the income security and safety nets to protect themselves from the health and economic damage caused by environmental degradation.

46 Desmond Tutu, 'We Do Not Need Climate Change Apartheid in Adaptation', in United Nations Development Programme, *Human Development Report 2007/2008: Fighting Climate Change*, New York: UNDP, 2007, 166; Harsha Walia, 'Against Eco-Apartheid: Towards Internationalist and Multi-Racial Worker Struggles', *Berliner Gazette*, 21 June 2023, berlinergazette.de.

47 See Rubah Salih and Olaf Corry, 'Displacing the Anthropocene: Colonisation, Extinction and the Unruliness of Palestine', *EPE: Nature and Space* 5(1), 2022, 381–400.

48 Ghada Sasa, 'Oppressive Pines: Uprooting Israeli Green Colonialism and Implanting Palestinian A'wna', *Politics* 43(2), 2023, 219–35.

49 Michael Quinlan, 'Precarity and Workplace Well-Being: A General Review', in Theo Nichols and David Walters (eds), *Safety or Profit? International Studies in Governance, Change and the Work Environment*, Abingdon-on-Thames: Routledge, 2014, 17–31.

Even if surplus workers are among the biggest losers of the ecological crisis, the immediate reaction to underemployment is often a demand for more capitalist growth. As seen in the introduction, productivity gains shorten the average labour time needed to make given quantities of material products. In other words, they reduce their value. As the productivity of labour increases, even modest growth in the amount of value produced by society accelerates the depletion of nature at a faster rate. The material throughput of society thus increases even in times of slow growth.[50] In fact, material output has been growing fast enough to ensure the deepening of the ecological crisis, but slow enough to keep precarity rising. Ilias Alami, Jack Copley and Alexis Moraitis have described the intractable combination of climate crisis, economic stagnation and surplus humanity as the 'wicked trinity' of capitalism.[51] The impression is that of a shift from the 1960s working-class refusal of work to a contemporary refusal of the working class *by* work. And yet, work is *absolutely* still looking for workers; it is refusing them only *relatively*. What is denied is the regularity of work and of the wages that come with it.

Dispossession by accumulation

Precarity is defined for our purposes as dependence on and irregularity in accessing a monetised income. Precarity, informality and service sector expansion (or tertiarisation) are mutually reinforcing phenomena, but they should not be conflated. Precarity leaks across formal and informal employment, and it is transversal to agriculture, industry and services. In addition, the surplus working class is itself internally

50 Nuno M.C. Machado, 'The Ecological Limit of Capitalism: Value-Form and the Accelerated Destruction of Nature in Light of the Theories of Karl Marx and Moishe Postone', in Vesna Stanković Pejnović (ed.), *Beyond Capitalism and Neoliberalism*, Belgrade: Institute for Political Studies, 2021, 111–22.

51 Ilias Alami, Jack Copley and Alexis Moraitis, 'The "Wicked Trinity" of Late Capitalism: Governing in an Era of Stagnation, Surplus Humanity, and Environmental Breakdown', *Geoforum* 153, 2023, 103691.

differentiated along many dimensions, including a continuum from the least secure to the least insecure employment arrangements. In his time, Marx had already noted the existence of different layers in the relative surplus population, and these stratifications have been taking manifold forms since then.[52]

As a rough simplification, it can be said that the global surplus working class is composed of all the dispossessed, whose reproduction is mainly tied to acquiring money, excluded from the 'standard' employment relationship. From this set, however, we must subtract middle-class professionals who might be engaged in non-standard employment, albeit with significant ownership or control functions (for example, a dependent self-employed business consultant).

Even under this simplification, it is difficult to put forward a quantitative estimate of the global surplus working class, as statistics on unemployment, informal employment and formal non-standard employment (such as short-term contracts, subcontracted labour, dependent self-employment, agency, part-time, on-call and 'gig' work) are uneven across countries. Furthermore, these statistics include both the remaining direct producers who still rely mostly on self-consumption, as well as some middle-class professionals, but exclude unwaged house workers, who are considered external to the active labour force.

Over the past two decades, the global unemployment rate has oscillated around 6 per cent, but this is not a very informative indicator.[53] On the one hand, the officially unemployed are only a subsection of the surplus working class. On the other hand, in countries where unemployment subsidies are minimal or non-existent, most people categorized as unemployed still have to work to survive.[54] 'Unemployment' is thus often a cultural construct signifying that somebody is engaged in a kind of work that is seen to be inadequate according to prevailing social norms. In any case, given that, as

52 Marx, *Capital*, vol. 1, 794–802.

53 Unless otherwise stated, the figures presented in this section are calculated based on the ILOSTAT Modelled Estimates repository, as well as on Tunisia's Institut National de la Statistique (INS) for data on Tunisia.

54 Aaron Benanav, 'The Origins of Informality: The ILO at the Limit of the Concept of Unemployment', *Journal of Global History* 14(1), 2019, 107–25.

of 2022, informal employment made up 58 per cent of the world's total employment, we can gauge that the surplus working class constitutes the vast majority of the global working class. In this sense, the 'standard' and 'non-standard' employment labels are counterintuitive, as the standard employment relationship is far from being 'normal'.

Furthermore, precarity should not be confused with poverty. The pre-capitalist peasantry, which accessed its means of reproduction predominantly through self-consumption, was poor and, in many ways, insecure, as it was highly exposed to the impacts of bad weather, diseases, famine, etc. The claim that capitalist development tends to increase insecurity in the broadest sense is for the most part false, at least for now. Nonetheless, the pre-capitalist peasantry was not precarious under the definition used here, as it did not depend on a monetised income and was thus shielded from the 'mute compulsion' of the law of value. The direct producers tied to the land are not part of the working class until their reproduction becomes subordinated in one way or another to money and the commodity form. When this happens, they might still remain in possession of some means of production, but not enough to cover most of their needs. They thus come to occupy a liminal position between the peasantry and the working class, whereby capitalist noxiousness contributes to impair their direct access to natural wealth, in what I have termed 'dispossession by accumulation'.

Owing mostly to China's development, the extreme poverty rate has decreased, along with the cost of consumer goods, from an estimated 38 per cent of the global population in 1990 to 8.4 per cent in 2019 (although the last years have seen a suspension of this trend).[55] The sheer amount of wealth produced by today's society allows most proletarians to toil or hustle their way to consumption in spite of high inequality. Despite its manifold irrationalities in both production and distribution, this abundance remains a pillar of enduring capitalist hegemony. However, given the productive powers of contemporary society, the persistence of absolute poverty has become even more scandalous. Furthermore, *absolute*

55 World Bank, *Poverty and Shared Prosperity 2024: Pathways Out of the Polycrisis*, Washington, DC: World Bank, 2024.

poverty reduction is not incompatible with *relative* immiseration. As ever more advanced technologies are applied to the labour process, wage goods become more affordable and are produced on an expanded scale. As noted in Chapter 1, real wages can rise as social needs are expanded and the cost of reproducing labour power is lowered, but this very cheapening shrinks the proportion of total output that returns to the workers. The accumulation of relative surplus value thus raises the rate of exploitation by shrinking relative wages (i.e. the labour share of income).[56]

The World Inequality Lab (WIL) has provided ample evidence of a long-term tendency towards rising income and wealth inequality between the early nineteenth century and today, but also of significant countertendencies. Some of the data presented might appear counter-intuitive to a Western public. Although the post–Second World War 'glorious thirty years' were a time of inequality reduction in the Global North, the report shows that global inequality reached its apex in 1980.[57] Global inequality has declined since then, accelerating its downward trend *after* the 2008 financial crisis (albeit remaining very high compared to 200 years ago). However, intra-country inequality grew significantly throughout the 1980–2020 period.[58] This means that global inequality reductions over the past four decades were mainly due to a decline in inter-country inequality, a rebalancing of economic growth between the Global North and some Global South countries, particularly China.

To some extent, this shift has altered the global balance of power established by European colonialism. However, it has little to do with a more egalitarian income distribution between labour and capital. According to the United Nations Conference on Trade and Development (UNCTAD), the global labour share of income in fact declined from 65

56 Rosa Luxemburg, 'Introduction to Political Economy' [1910], in *The Complete Works of Rosa Luxemburg*, vol. 1, *Economic Writings 1*, ed. Peter Hudis, London: Verso, 2014, 89–300, 286.

57 Lucas Chancel et al., *World Inequality Report 2022*, Paris: World Inequality Lab, 2022, 55.

58 Ibid., 57.

to 59 per cent of GDP between 1980 and 2023.[59] These estimates are highly conservative, because the labour share that they indicate includes the waged incomes of capitalists (such as CEOs), which have been rising.

For decades, a broad range of mobilisations, both in the Global North and in the Global South, have been denouncing precarity in order to prop up working-class rigidity against the flexibility demanded by the law of value. Despite this, a denialist literature on precarity has surfaced within some strands of Marxist theory. These authors usually wear their arguments under a cloak of orthodoxy, claiming that, to the extent that it exists, precarity is a constant element of capitalism. However, Marx argued that there is a tendency for the rate of the surplus working class to rise. It is true that precarity has always existed, yet – according to Marx – precarity is not a static element of capitalism, but one tending to expand over time: 'The working population . . . produces both the accumulation of capital and the means by which it is itself made relatively superfluous; and it does this to an extent which is *always increasing*.'[60]

Nonetheless, while the tendency towards precarity rising exists *in theory*, it is not straightforward to verify *empirically* whether it came to fruition, as countertendencies could have prevailed. In this respect, even the denialist literature usually recognises that the post-war trend towards expanding job security in the Global North is no longer in place. However, suggestions that we have merely returned to a pre-Fordist capitalist 'normality', or that nothing has changed in the Global South, are misleading.

First, we are nowhere near the nineteenth-century situation in which most of the global direct producers relied on self-consumption to meet their needs, and freshly dispossessed peasants-in-transition had the freedom to migrate to 'new worlds'. Precarity has risen secularly in the

59 United Nations Conference on Trade and Development, *Trade and Development Report 2023: Growth, Debt, and Climate, Realigning the Global Financial Architecture*, New York: United Nations, 2023, 15.

60 Marx, *Capital*, vol. 1, 783, emphasis added.

sense that gradual proletarianisation has made the global population more dependent on obtaining a regular monetary income for its normal reproduction, while denying such a regular income to most.

Second, the Global South is not an undifferentiated and static entity. Conservative regimes such as Morocco never really challenged the dominance of precarious employment.[61] However, between the 1930s and the 1970s, in many other countries, developmentalist or 'socialist' states attempted to expand industry through import-substitution industrialisation. While most of their working classes remained precarious, a trajectory to expand secure employment through the public sector was established, as seen in 'developmentalist' social pacts in Latin American and African countries, or the Chinese 'iron rice bowl' system.[62] This trajectory towards increasing job security for broadening segments of the workforce was interrupted by the neoliberal turn of the 1970s and '80s, and the shift towards privatisation and austerity. While labour law reforms weakened guarantees for permanent employees, the neoliberal phase also saw a proliferation of formalised precarious employment arrangements. This led the ILO to state that 'over the past few decades, in both industrialized and developing countries, there has been a marked shift away from standard employment to non-standard employment'.[63]

Again, the underlying driver of rising precarity is the general law of capitalist accumulation. A crucial effect of the pincer movement driven by capitalist automation was the secular decline in the global share of

61 Lorenzo Feltrin, 'Labour and Democracy in the Maghreb: The Moroccan and Tunisian Trade Unions in the 2011 Arab Uprisings', *Economic and Industrial Democracy* 40(1), 2019, 42–64.

62 Hishaam D. Aidi, *Redeploying the State: Corporatism, Neoliberalism, and Coalition Politics*, London: Palgrave Macmillan, 2009; Margarita Fajardo, *The World That Latin America Created: The United Nations Economic Commission for Latin America in the Development Era*, Cambridge, MA: Harvard University Press, 2022; Ching K. Lee, 'Precarization or Empowerment? Reflections on Recent Labor Unrest in China', *Journal of Asian Studies* 75(2), 2016, 317–33.

63 International Labour Organization, *Non-Standard Employment Around the World: Understanding Challenges, Shaping Prospects*, Geneva: International Labour Office, 2016, 2.

agricultural employment. Up to a point, agricultural employment losses were offset by the increase in the share of industrial jobs. Yet, as we have seen, the global share of industrial employment has been declining at least since the early 1990s. Over recent decades, the service sector has therefore borne the burden of absorbing workers evacuated from both agriculture and industry. In many Global South countries, which never industrialised comparably to the West, workers are being thrown directly from agriculture into services, or kept in a limbo between the two. For the vast majority of them, service sector employment does not mean a handsomely paid job in finance or creative work in the media industry, but insecure toil in low-end, slow-growth services.[64] Therefore, there is no universal and linear three-stage development model, in which all countries necessarily graduate from agriculture to industry and then finally to services. In a world market saturated with cheap industrial goods, very late industrialisers may well turn out not to be industrialisers at all. They may never reach the share of industrial employment obtained by Western countries at their peak.

The transformations of the Tunisian class composition are an example of how the global tendencies just described can be instantiated at a national level. Between 1961 and 2014, officially recorded agricultural employment in Tunisia fell from 45 to 15 per cent of total employment (Figure 2.1). This is the result of pressures on the peasantry that reach back to colonisation, manifest in land grabbing, landholding de-collectivisation and fragmentation, agricultural modernisation and the rise of export monocrops.[65] In addition, the ecological crisis has been acting as a driver of dispossession by accumulation, as water scarcity, desertification, extreme weather events, plagues and wildfires make it harder for rural producers to live off their land.[66]

64 See Ricardo Antunes, *The Meanings of Work: Essay on the Affirmation and Negation of Work*, Leiden: Brill, 2013 [1999].

65 Habib Ayeb and Ray Bush, *Food Insecurity and Revolution in the Middle East and North Africa: Agrarian Questions in Egypt and Tunisia*, London: Anthem Press, 2019.

66 Karolina Sobczak-Szelc and Naima Fekih, 'Migration as One of Several Adaptation Strategies for Environmental Limitations in Tunisia: Evidence from El Faouar', *Comparative Migration Studies* 8(8), 2020.

Up to the early 1980s, Tunisia's industry partially compensated for the secular decline in agricultural employment. Industrial employment (in the strict sense) rose from 7 per cent in 1961 to 19 per cent in the mid-1980s (Figure 2.1).[67] It subsequently stagnated between 18 and 20 per cent but changed qualitatively. Relatively secure employment in state-owned enterprises (SOEs) declined, while precarious employment in private-sector light manufacturing for export increased. Tunisia too has its Rustbelts and Sunbelts. The state remains involved in some basic industries, some of which are highly polluting, like phosphate mining and processing.[68]

Meanwhile, public administration employment grew from 13 per cent of total employment in 1961 to 17 to 18 per cent in the mid-1980s, after which it remained at the same level until the 2011 uprising. This means that, since the 1980s, the burden of absorbing workers expelled from the land has fallen entirely on the shoulders of a *private* service sector characterised by high insecurity. In fact, the share of employment in services other than public administration increased from 18 per cent in 1961 to 28 per cent in 2014. In parallel, informal employment was estimated to stand at 44.8 per cent of Tunisia's total employment in 2019.

The Tunisian workforce also follows the global tendencies towards rising educational levels, increasing female participation in official employment, and significant mobility. Internal migrations have been historically significant, mostly from the interior to the coastal regions, where many migrants are employed in light manufacturing and tourism. While large numbers of Tunisian workers have been emigrating over recent decades, immigration from abroad used to be a relatively small phenomenon. Since the 2010s, however, growing numbers of sub-Saharan Africans have got stuck in Tunisia on their way to Europe, which has engendered a wave of racial discrimination and tensions in the

67 These figures were calculated using the database created by Monji Ben Chaabane for his study *La rémunération des salariés, 1989–2014*, Tunis: Institut Tunisienne de la Compétitivité et des Études Quantitatives, 2014, itceq.tn.

68 Habib Ayeb, 'The Marginalization of the Small Peasantry: Egypt and Tunisia', in Habib Ayeb and Ray Bush (eds), *Marginality and Exclusion in Egypt*, London: Zed Books, 2012, 72–96.

Figure 2.1 Tunisia's total employment by economic sectors

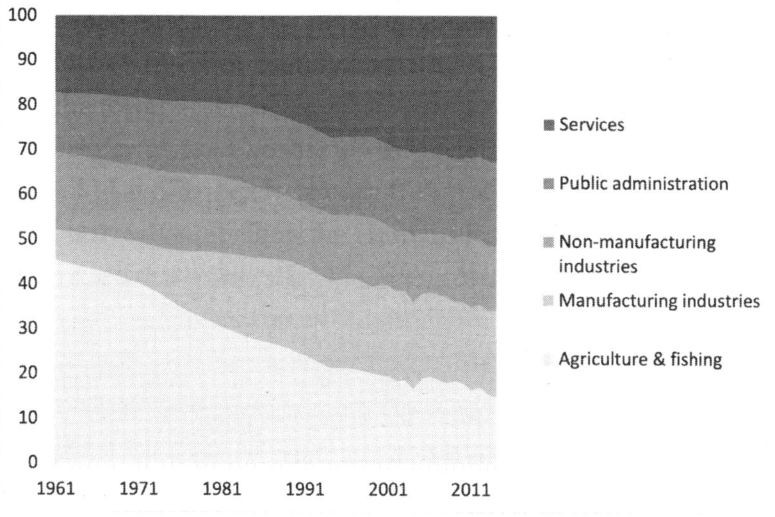

Source: Ben Chaabane, *La rémunération des salariés, 1989–2014.*

country.[69] These migrants mostly come from Western and Central Africa, which are among the world regions worst affected by climate change.[70]

These changes in Tunisia's technical working-class composition were not a natural and conflict-free evolution. After Independence from the French was won in 1956, Tunisia engaged in an import-substitution industrialisation programme with the creation of several heavy-industry SOEs, while the traditional agricultural sector was modernised through a system of state-run 'cooperatives'. A new cycle of struggles then followed, spanning from the start of a strike wave in 1971 to the repression, in 1985, of the main labour confederation, Union Générale Tunisienne du Travail (Tunisian General Labour Union, UGTT).[71] While precarious workers also mobilised, the protagonists of this cycle were a

69 Riadh Ben Khalifa and Mehdi Mabrouk, 'Discours sur l'immigration subsaharienne en Tunisie: Le grand malentendu!', *Confluences Méditerranée* 2(125), 2023, 91–106.

70 International Labour Organization, *Working on a Warmer Planet*, 18.

71 Hèla Yousfi, *Trade Unions and Arab Revolutions: The Tunisian Case of UGTT*, Abingdon-on-Thames: Routledge, 2017 [2015].

new generation of educated workers and waged middle-class youths who had been radicalised in the student movement before entering both the workforce and the UGTT.[72] These workers, supported by the UGTT left, and in many cases against the UGTT National Executive Committee, led the strike wave that peaked in 1977. The strikes subsided in 1978–80 due to the harsh repression that followed the general strike of 26 January 1978, but they picked up again in the early 1980s. The strongholds of industrial militancy were large SOEs, while strikes in transport were also widespread.[73] The core demands revolved around the wage, and real wages did grow quickly, surpassing the GDP growth rate.[74]

The regime thus made generous concessions at first, but rising debt and the lack of hard currency pushed it towards a typical structural adjustment programme under the supervision of the International Monetary Fund (IMF) and the World Bank. In 1985, the authorities escalated their repression of the UGTT, which included attacks by unofficial militias on union premises, while many rank-and-file union activists were laid off and senior UGTT leaders were jailed.[75] Following a common pattern, the former strongholds of labour militancy were restructured, privatised or abandoned to slow decline. Total public employment (the public administration plus the SOEs) declined from 41 per cent of wage-earners in 1980 to 32 per cent in 2010.[76] Yet employment in the public sector is still seen as the way to economic security and upward social mobility. Times of weak investment and fiscal crisis

72 Salah Hamzaoui, 'Quelques réflexions et hypothèses pour l'étude de la jeunesse ouvrière', *Cahier du CERES, Sociologie* 10, 1984, 257–65.

73 Salah Zeghidi, 'L'UGTT: Pôle central de la contestation sociale et politique', in Mahmoud Ben Romdhane (ed.), *Tunisie: Mouvements sociaux et modernité*, Paris: Karthala, 1997, 13–62.

74 Mahmoud Ben Romdhane, *Tunisie. État, économie et société*, Paris: Publisud, 2011.

75 Habib Achour, *Ma vie politique et syndical. Enthousiasme et déceptions (1944–1981)*, Lyon: Alif, 1989.

76 International Monetary Fund, *Tunisia*, IMF Country Report No. 14/277, Washington, DC: IMF, 2014; Abderrazak Zouari, 'Collective Action and Governance Structure in Tunisia's Labor Organization', in Mustapha K. Nabli and Jeffrey B. Nugent (eds), *The New Institutional Economics and Development: Theory and Applications to Tunisia*, Amsterdam: North Holland, 1989, 323–51, 339.

have only exacerbated the oversubscription of this sector of the labour market.

The latest major cycle of struggle can be plotted from the beginning of the Gafsa Revolt for secure employment and local development in January 2008 to President Kais Saied's July 2021 takeover.[77] Instead of relatively secure waged workers, the protagonists in these struggles were the precarious youth. While the UGTT often held solidarity strikes with their protests, the precarious workers' very conditions made it impossible for them to be union members and to use strikes as a weapon. The majority did not belong to any stable, formal organisation. While broadly left-wing ideas and sensibilities prevailed in the 1970s, the Tunisian left of the 2010s had to compete with a strong Islamist movement, as well as widespread 'anti-politics' feelings. Visitors who hoped to be welcomed by red stars and banners were disappointed. One could see them everywhere, but only on the Tunisian national flag.

Roadblocks and riots were the Tunisian precarious workers' main tools for seeking concessions. Less sensational than riots, roadblocks can produce greater economic damage and have the advantages of being more sustainable and risking less harm to the participants and others. Because they halt production from outside the workplace, roadblocks are effectively strikes by the 'workless' working class. The 'workless' can exploit logistical weak spots to block the circulation of strategic commodities and slow down or stop production as a consequence.[78] Unlike strikes, however, roadblocks are illegal, and therefore their practicability depends on the balance of power between protesters and law enforcement. Some of these protests also had an explicit ecological dimension. Most notably, the roadblocks for secure employment at the Gabes phosphate-processing complex, which discharges toxic phosphogypsum

77 Amin Allal, 'Réformes néolibérales, clientélismes et protestations en situation autoritaire: Les mouvements contestataires dans le bassin minier de Gafsa en Tunisie (2008)', *Politique africaine* 117, 2010, 107–25; Larbi Chouikha and Vincent Geisser, 'Retour sur la révolte du bassin minier: Les cinq leçons politiques d'un conflit social inédit', *L'année du Maghreb* 6, 2010, 415–26.

78 See, e.g., Max Ajl, 'Development by Popular Protection and Tunisia: The Case of Tataouine', *Globalizations* 16(7), 2019, 1215–31.

directly into the sea, intersected with a sustained movement for health and environmental protection.[79]

The famous case of Mohamed Bouazizi, whose self-immolation triggered the 2011 uprising, is emblematic. Bouazizi was an agricultural labourer who lost his job after the land on which he worked was sold and production there restructured.[80] The day Bouazizi set himself ablaze, he had been harassed by the police while illegally selling vegetables on the street. Bouazizi's trajectory can thus be read as an illustration of the general trend described above, in which losses in agricultural employment are not compensated by new jobs elsewhere, forcing many young people into exasperating precarious work particularly in the low-end of the service sector. To survive, the precarious have to resort to individual hustling or collective mobilisation. In January 2011, collective mobilisation met a rigidly authoritarian state-form and managed to bring it down as the precarious were joined by other sectors of the population, including many secure workers and members of the middle class. However, this cycle of struggles could not consolidate the political and socioeconomic concessions it initially achieved. Rather, the long-drawn-out crisis that accompanied the uprising, and which has continued into the present, has had the effect of deepening precarity further.

Kerkennah, Tunisia: Oil, gas and blue crabs

The Kerkennah archipelago, in the Gabes Gulf, is at the frontline of climate change due to its vulnerability to rising sea levels, which have accelerated coastal erosion and soil salinisation.[81] This process is

79 Diane Robert, 'Protest Movements Against Industry-Related Environmental Burdens and Territorial Injustice in Gabès and Kerkennah (Tunisia)', *Spatial Justice* 16, 2021.

80 Mathilde Fautras, 'Land Injustices, Contestations and Community Protest in the Rural Areas of Sidi Bouzid (Tunisia): The Roots of the "Revolution"?', *Spatial Justice* 7, 2015.

81 Hamza Hamouchene, 'Tunisia: On the Frontlines of the Struggle Against Climate Change', *ROAR Magazine*, 28 July 2016, roarmag.org.

aggravating freshwater scarcity on the islands and expanding *sebkhas*, sterile salt flats where palms and other vegetation no longer grow.[82] Arriving in Kerkennah from the mainland port of Sfax, visitors cannot fail to notice the poor state of the infrastructure, except for the main road connecting the south-western port of Sidi Youssef to the north-eastern village of Alataya. That was the road used by Petrofac trucks to transport gas condensate out of the islands. Yet, Kerkennah is also one of the main departure points for both Tunisian and sub-Saharan illegal emigration journeys to Europe. The Italian island of Lampedusa is only 150 km away and many young Kerkennians have tried their luck on the dangerous trip. Lethal shipwrecks between Kerkennah and Lampedusa are appallingly frequent. At least 112 people lost their lives on 3 June 2018, while sixty-one people met the same fate on 5 June 2020, and thirty-nine people on 9 March 2021, to mention just the most severe episodes.

In 2016, Kerkennah had a resident population of only about 15,500, with an average age of thirty-seven – five years above the national average due to its high emigration rates. The inhabitants are scattered among thirteen villages, the largest being Mellita, Alataya and the capital Remla. Its class composition is partially distinct from the national picture because a large proportion of the employed (42 per cent) work in the primary sector, due to the dominance of fishing in the local economy. Women make up 21.5 per cent of the official economically active population, but this figure hides unwaged housework, small subsistence farming and activities auxiliary to fishing, like the production or maintenance of fishing nets. Kerkennah's self-employed fishers have long been integrated in the market economy. In fact, they sell their product to wholesale and food-processing companies who thereby appropriate part of the value they produce. What Marx wrote on agricultural smallholders applies here: 'The smallholding of the peasant is only a means for capitalists to draw profit, interest, and rent from the soil, leaving to the

82 Noômène Fehri, 'La palmeraie des Îles Kerkennah (Tunisie), un paysage d'oasis maritime en dégradation: Déterminisme naturel ou responsabilité anthropique?', *Physio-Géo* 5, 2011, 167–89.

farmer himself how to extract his wages.'[83] Nevertheless, because a strong element of self-consumption remains regarding the fishers' food needs, they are best seen as semi-proletarians, combining dependent self-employment for money with non-monetised access to wealth from the marine commons.

Artisanal fishing in Kerkennah is centred on *charfia*, a tradition dating back to the Phoenician era. *Charfia* deploys mazes of pathways and 'chambers' built into the sea with palm leaves, which lure the fish into palm-fibre traps (*drina*).[84] Another technique consists in ensnaring octopuses in clay amphoras (*qaroor*). Artisanal fishing has, however, increasingly suffered from environmental degradation, as the sea is more and more polluted with hydrocarbons and metal-bearing microplastics.[85] In addition, rising sea temperatures have led to the proliferation of 'invasive' species, particularly spider and blue crabs.[86] Due to the 'terrorist' inclinations of these pincered decapods, the local fishers call them *daesh*, Arab for ISIL (Islamic State of Iraq and Levant). The crabs prey on indigenous marine species, damage the fishing nets to pillage the catch inside and even pinch the fishers' fingers, becoming a real occupational hazard. A survey estimated that the crab 'invasion' had decimated Gabes Gulf fishers' income by 72 per cent.[87] Kerkennah's fishers have thus been

83 Karl Marx, 'The Eighteenth Brumaire of Louis Bonaparte' [1852], in *Marx: Later Political Writings*, ed. Terrell Carver, Cambridge: Cambridge University Press, 1996, 31–127, 120.

84 Moussa Hamdi, Abdessalem Shili and Sonia Nasraoui, 'The Traditional Fishery "Charfia" in Chebba (Middle Eastern Tunisia): Technical Characteristics, Catch Composition and Related Social Issues', *Bulletin de l'Institut National des Sciences et Technologies de la Mer de Salammbô* 46, 2019, 41–50.

85 Hatem Zaghden et al., 'Origin and Distribution of Hydrocarbons and Organic Matter in the Surficial Sediments of the Sfax-Kerkennah Channel (Tunisia, Southern Mediterranean Sea)', *Marine Pollution Bulletin* 117, 2017, 414–28.

86 Guillame Marchessaux et al., 'Invasive Blue Crabs and Small-Scale Fisheries in the Mediterranean Sea: Local Ecological Knowledge, Impacts and Future Management', *Marine Policy* 148, 2023, 105461.

87 Wafa Rjiba-Bahri et al., 'Morphological and Biological Traits, Exoskeleton Biochemistry and Socio-Economic Impacts of the Alien Invasive Crab *Libinia dubia* H. Milne Edwards, 1834 from the Tunisian Coast (Central Mediterranean)', *Thalassas* 35, 2019, 291–303.

suffering from the dispossession by accumulation engendered by the ecological crisis.

These difficulties have encouraged several fishers to abandon *charfia* for the more lucrative, albeit banned, method of trawling called *kiss* (Arab for 'bag'). Much of the illegal catch is then exported to Italy and Spain via European wholesalers.[88] The *kiss* system can therefore be seen as European capital saving costs by outsourcing destructive fishing methods to Tunisian boat owners. And it is a process that has generated a vicious circle. The more people trawl, the more they undermine the bases for sustainable fishing by damaging the seabed and its carbon-absorbing Neptune seagrass meadows (also known as 'the lungs of the Mediterranean').[89] This, in turn, pushes even more fishers to embrace *kiss* to make a living. Another result of these pressures has been the hybridisation between artisanal and illegal fishing. *Charfia*'s handicraft implements are now often combined with forbidden industrial ones, such as polyvinyl chloride (PVC) stakes and nylon gillnets, which further contribute to marine plastic pollution.[90]

Alongside its struggle with environmental degradation, fishing on the islands has also had to co-exist with fossil fuel extraction since the 1990s. Originally, the two main energy companies operating in Kerkennah were Thyna Petroleum Services (TPS) and Petrofac. The latter, however, was replaced by Perenco in 2018. Petrofac is a Jersey-registered corporation, convicted by the UK Serious Fraud Office (SFO) for 'seven separate counts of failure to prevent bribery between 2011 and 2017'.[91] The

88 Davide Mancini, Sara Manisera and Arianna Poletti, 'How Illegally Caught Fish in the Mediterranean Enter Europe', *Geographical*, 1 March 2023, geographical.co.uk.

89 Radhouan El Zrelli et al., 'Economic Impact of Human-Induced Shrinkage of Posidonia Oceanica Meadows on Coastal Fisheries in the Gabes Gulf (Tunisia, Southern Mediterranean Sea)', *Marine Pollution Bulletin* 155, 2020, 111124.

90 Khawla Chouchene, Teresa Rocha-Santos and Mohamed Ksibi, 'Types, Occurrence, and Distribution of Microplastics and Metals Contamination in Sediments from South West of Kerkennah Archipelago, Tunisia', *Environmental Science and Pollution Research* 28, 2021, 46477–87.

91 'Serious Fraud Office Secures Third Set of Petrofac Bribery Convictions', Serious Fraud Office, 4 October 2021, sfo.gov.uk.

corporation started extracting gas in Kerkennah's Chergui field in 2008. Unlike fishing, the capital-intensive gas industry provided only a very limited number of jobs in Kerkennah, about eighty direct ones according to the local union, in addition to fifty-five security workers hired through a local agency.

The relative poverty of the islands compared to the mass of wealth extracted by rich foreign multinationals generates strong feelings of injustice among the population. I often heard complaints that the vast majority of Kerkennians have never benefited from these extractive activities, even if they pollute the environment and damage fishing and tourism. Ahmed Souissi, a long-time left-wing activist and the spokesperson for Union des Diplômés Chômeurs (Union of the Unemployed Graduates, UDC) at the time, told me: 'The demands were clear: development and employment for the islands. "You're exploiting a great oil and gas wealth, in return you pollute the islands, the sea where we fish . . . and we don't see any benefits."' These demands, then, were not strictly speaking 'anti-extractivist' as they did not call for an end to fossil fuel exploitation. However, they did point towards enhanced community control over its territory, aiming to put a brake on the vicious circle of environmental degradation, precarity and emigration that had set in on the archipelago.

Such discontent engendered a social movement whose class composition can be broken down into four main segments: unemployed graduates, fishers, public administration employees and Petrofac workers (direct and indirect). The unemployed graduates were the most prominent group mobilising against Petrofac. Most male unemployed graduates in Kerkennah actually worked as fishers, just like many other men on the islands. As the Tunisian private sector offers few jobs for people with educational qualifications relative to the supply of graduates, they demanded employment in the public sector. Both women and men participated, and women made up the majority of the 266 participants in the Petrofac-funded employment scheme (see below). During the height of the mobilisations, the movement organised through village-level assemblies and delegates, and a general assembly in Remla. The UDC, the organisation that represents unemployed graduates, has a leftist national leadership, while its membership is open to all unemployed people.[92]

Just like the unemployed graduates, the fishers are not unified by the condition of being waged employees working in large groups under the same employers. But this lack of occupational cohesion is compensated within each category and between them by strong communal ties, as many unemployed graduates also fish, and all have fishers in their families. Virtually all men on the islands are able to fish and do so more or less regularly depending on the extent to which they rely on it for their income. Small-scale fishing is a hard, dangerous and insecure job. Fishers either own their boats or share their catch with the owners of the boats they use. The 'boatless' fishers are paid in cash by the boat owner for a quota of the product. The boat owner then sells all the catch to local wholesalers, who supply national and European retail markets and food-processing factories.

The fishers did not benefit from fossil fuel extraction, and TPS's recurrent oil spills were highly detrimental to their livelihood. Indeed, the first protests against the oil and gas companies were led by fishers demanding compensation for the damage caused by oil leaks. Most fishers agreed that the energy firms should contribute to the development of the islands and have therefore often participated in UDC-led mobilisations. Mellita has a fishers' union affiliated to the UGTT, but most fishers tend to organise on an informal basis.

Public administration employees are the other large occupational group in Kerkennah, making up 23 per cent of the employed population and forming the backbone of the UGTT. Despite an economy based on artisanal fishing, trade unionism has had an important role in the history of the islands. In fact, the UGTT's main founders, Farhat Hached and Habib Achour, were Kerkennians who had migrated to the mainland, while several UGTT national leaders since have had Kerkennian origins, maintaining the union's ties to the archipelago.

92 See Samiha Hamdi and Irene Weipert-Fenner, 'Unemployed Protests in Tunisia: Between Grassroots Activism and Formal Organization', in Irene Weipert-Fenner and Jonas Wolff (eds), *Socioeconomic Protests in MENA and Latin America: Egypt and Tunisia in Interregional Comparison*, London: Palgrave Macmillan, 2020, 195–219.

While public administration employees are certainly better off than fishers and unemployed graduates, they are still linked to these other segments by family and community bonds, and they did not benefit from Petrofac's presence either. Public administration employees were therefore mostly in agreement that the corporations should contribute to local employment and development.

The Petrofac employees and security workers depended on the corporation's presence on the islands for their jobs, which offered better wages and conditions than they could get elsewhere. Petrofac also had an arrangement with Kerkennah's Grand Hotel that helped to keep it afloat during the low season. These workers therefore faced strong contradictory pressures. On the one hand, there were Petrofac's threats to leave Kerkennah. On the other, there were their own grievances against Petrofac and feelings of solidarity towards their fellow islanders.

The local UGTT was therefore 'between a rock and a hard place'. It had to respond to anti-Petrofac pressures from its base in the public administration and from the general population, while still defending the interests of its members who depended on Petrofac for their work. The union thus mediated between the Petrofac workers and the precarious. Its position was that Petrofac should continue operating in Kerkennah, but it backed the unemployed graduates' demands. Eventually, UGTT-affiliated Petrofac employees, along with security workers and staff from the Grand Hotel, joined the local general strike against repression.

The Kerkennah movement against Petrofac originated in the 2011 Tunisian Uprising. Petrofac had been targeted by sporadic protests since its early years in the archipelago, but only with the uprising was it really forced to take the population's grievances seriously. Kerkennah joined the uprising on 12 January 2011, with the regional general strike called by the Regional Executive Committee of the Sfax UGTT. On 14 January, clashes between protesters and police left twenty-six-year-old demonstrator Slim Hadhri dead. In the following weeks, protesters held sit-ins outside the local government offices and at the Petrofac field, blocking the trucks transporting gas to the Sidi Youssef port. In the post-uprising context of political instability and widespread social mobilisation, the authorities and Petrofac itself were willing to make concessions to keep the situation under control. On 20 May 2011, an agreement was signed according to which Petrofac would provide 600,000 Tunisian dinars (TDNs) per year (€303,000 at the time) to boost employment and development on the islands.

Following the agreement, an employment scheme was put in place in which Petrofac disbursed the funding to the governorate, which then forwarded it to 248 unemployed graduates affiliated to the UDC. Their number increased to 266 in 2012, 153 of them women, with the unemployed graduates being allocated to different public administration agencies on the islands. It was clear, however, that both the authorities and Petrofac saw this as a temporary arrangement. There were no contracts, no payslips and no social security contributions. Indeed, the regularisation of their employment then became the unemployed graduates' main demand, generating periodic protests.

In March 2015, Petrofac declared its intention to halt the employment scheme. In response, the unemployed graduates set up a new roadblock, stopping production on the site from 10 March until 16 April, when a new agreement between the UDC, the UGTT and Petrofac was signed. The company agreed to keep financing the employment scheme until the end of 2015. In the meantime, the governorate would establish a firm for the provision of 'environmental' services, modelled on the employment schemes created after the uprising in other marginalised regions. In January 2016, however, the authorities did not implement the agreement and the conflict thus turned into a deadlock.

On 19 January 2016, the unemployed graduates started a new blockade of the Petrofac field and halted production once again. This time, however, the authorities and Petrofac were not willing to back down, as indicated by the gangs of thugs that unsuccessfully attacked the protesters. A female activist mentioned that: 'Everybody [taking part in the movement] was harassed but there were attempts to insult and terrorise the women specifically, although they resisted well.' On 3 April, word spread that riot police were aboard the public transport ferry from Sfax to Kerkennah. During the night, at least 500 policemen attacked the roadblock with tear gas and water cannons. This massive display of force was taken as a provocation by many inhabitants, especially the fishers, who joined the fray even if they had not been part of the original blockade.

The roadblock reassembled soon after its dispersal, while the riot police maintained their presence on the islands and made four arrests. This stasis continued for several days, until 8 April, when the UGTT declared a local general strike to be held on 12 April. The strike, which called for the liberation of the arrested and the resumption of negotiations to satisfy the movement's demands, was successful, with the participation of much of the population as well as local businesses. On 14 April, Mohamed Ali Arous, the Kerkennah local UGTT secretary general, secured a twenty-four-hour 'truce' in order to propose an agreement to the protesters. However, the police did not respect the truce and, on the same evening, used force to get the Petrofac trucks through from the port of Sidi Youssef to the gas field. Tension had mounted over the previous ten days and now exploded. Riots against the police broke out all over Kerkennah. On the same night, protesters at the mainland port of Sfax set up a roadblock to prevent police reinforcements from leaving for the islands. On 15 April, the clashes picked up again until the police were cornered in the port of Sidi Youssef with no escape route. The protesters organised to let the officers board the boat back to Sfax without anybody getting seriously hurt.

After the April crisis, the situation returned to deadlock. The police were not allowed back on the archipelago, Petrofac remained closed even in the absence of a physical roadblock, and no negotiations took place. Eventually, in September 2016, the government brokered a new agreement with larger concessions than the earlier ones, including a fund worth TDN 3.5 million per year (about €1.4 million at the time) for local development projects, with a local committee created to decide on the allocation of these funds. All charges against the protesters were to be dropped and the energy firms would be allowed to operate normally. In the following days, law enforcement returned to Kerkennah and Petrofac resumed production after a blockade that had lasted eight months.

This did not last long. In the summer of 2017, Petrofac abandoned its Kerkennah field to the Anglo-French corporation Perenco, so that the agreement had to be re-negotiated again. While the movement was able to win many concessions, these were insufficient to address the downward spiral of employment precarity and environmental degradation that is affecting the region more generally.

This story is, in several respects, representative of global trends: the most productive sectors of the Kerkennian economy are highly capital intensive and thus require a tiny number of workers, and a large part of the population lives off various kinds of precarious employment while being mostly excluded from the gains generated by high-tech production. Moreover, the archipelago's protest dynamics are consistent with the feminist insight that the realm of workers' reproduction, the community, can become a site of struggle hindering the accumulation of capital. The people who took part in the mobilisations were relatively fragmented on an occupational level, with no large groups working for the same employers. However, the pressures towards class decomposition generated by employment precarity and dispersion were more than counterbalanced by a recomposition centred on communal solidarity. As UDC's Ahmed Souissi commented: 'Being the inhabitants of an island, with the boats connecting us to the mainland only between 6 am and 6.30 pm, and otherwise we're isolated from the rest of the world . . . this creates a kind of special mentality. We told the prefect and the governor never to consider the use of force, because it would turn the whole island upside down . . . Even if we were only about 150 people, far away from the villages, at the Petrofac field. Even those Kerkennians who were openly against us, like Petrofac workers or some hotel owners, they will switch to our side to defend the island.'

Kerkennah also exemplifies how the pincer movement of capital accumulation tends to dispossess the direct producers of their livelihoods with no guarantee of secure employment in the money economy. These surplus workers are, to varying extents, refused by both the land (or the sea, as in this case) and regular work simultaneously. Dispossession by accumulation, whereby capitalist noxiousness accelerates the separation of the producers from nature through environmental degradation, is also a reality for Kerkennah's fishers. However, while in the stagist schema peasants-in-transition moved from a growingly commodified agricultural sector into the factories, a more common scenario is that described by the Latin American literature on marginality: a transition from the land to little more than nothing, to a limbo between the subsistence economy and the surplus working class.

These trends are particularly accentuated in the Middle East and North Africa, as this region as a whole is disproportionately dependent on the export of rent-bearing primary products, especially fossil fuels. Hydrocarbons are both notoriously destructive for the environment and prone to produce large surplus populations. In fact, the valorisation of fossil-extracting capital depends relatively little on the exploitation of its own employees, because rents are appropriated from the global working class through high prices. The oil-rich Gulf monarchies have been siphoning off such rents towards supporting the repression of surplus working-class organising across the region, or the penetration of communal solidarity by the Islamist right, exacerbating the sectarian divisions into which social anger can be channelled. If we are to better understand the globally uneven distribution of precarity and noxiousness that makes this possible, then we need an analysis of the international division of labour, to which the next chapter turns.

3

The International Division of Labour and Noxiousness: The Mutations of the Ecological Transition from Above

In 1966, the Soviet government and the Italian automotive company FIAT agreed on a partnership to establish the largest car factory in the country of 'actually existing socialism'. To house the workforce of what would become AvtoVaz, the planners chose Tolyatti, the town on the Volga named after former PCI leader Palmiro Togliatti. The documentary *Togliatti(grad)* tells this story through the voices of workers and managers. Russians were shipped to Turin to learn the labour process in FIAT's own factories. Landing in Italy in the middle of the Hot Autumn of 1969, however, they could hardly get anything done, as they were continuously stopped at the factory gates by strike pickets. In the other direction, Italians were flown to Tolyatti to transfer the knowledge necessary to operate the second-hand machinery that had found a new life there. These FIAT employees were mostly communist technicians volunteering to contribute to Soviet industrialisation. When the Russian workers went on strike against the brutal health and safety conditions of AvtoVaz's early days, the Italian communists decided to join them in solidarity. They were, however, prevented from doing so because they had agreed to respect Soviet laws, including restrictive strike regulations, before leaving for Russia. In the end, the Russians eager to work in Italy were prevented from doing so, while the Italians ready to strike in the Soviet Union had to keep working. This paradox, one may conclude, reflects the contradictions of challenging the international division of labour through the productivist adoption of capitalist technologies, including a mobility system centred on private cars.

In 1975, a few years after the establishment of AvtoVaz, *operaista* intellectual Luciano Ferrari Bravo published the essay 'Old and New Questions in the Theory of Imperialism' as an introduction to the edited volume *Imperialism and the Multinational Working Class*.[1] As Sergio Bologna recalls, the political context in which Ferrari Bravo wrote this article was marked by contentions between *operaistas* and Third Worldists within the Italian radical left:

> The *operaistas* ceaselessly denounced the opportunism and sterility of the positions of those who exalted the 'jungle guerrillas' of the Third World but remained indifferent to the struggles in countries where they lived or idealised them as the 'asphalt guerrillas' of the capitalist core (thus creating the premises for some of the terrorist practices of the following years).[2]

Nonetheless, after a long review of the debates on imperialism, Ferrari Bravo singled out the Third Worldist Arghiri Emmanuel, while disagreeing with his political conclusions, as the most compelling author on the topic.[3] This unlikely appreciation is explained by one element: the wage. More precisely, the role of the wage in the international division of labour. To gain a fuller understanding of the latter, however, we need to consider the relationship between the wage and its antagonists: profit and rent. With this objective, the first section of this chapter builds a dialogue between *operaismo* on the one hand and dependency theory

1 Luciano Ferrari Bravo (ed.), *Imperialismo e classe operaia multinazionale*, Milan: Feltrinelli, 1975.

2 Sergio Bologna, 'Prefazione', in Luciano Ferrari Bravo, *Dal fordismo alla globalizzazione: Cristalli di tempo politico*, Rome: Manifesto Libri, 2001, 7–35, 14.

3 Luciano Ferrari Bravo, 'Old and New Questions in the Theory of Imperialism' [1975], *Viewpoint Magazine*, 1 February 2018, viewpointmag.com.

and *iñiguismo* on the other, the latter being two Marxist traditions developed mainly in Latin America.

This framework is then deployed to dissect recent ecological transition attempts, and the related transformations in the international division of labour and noxiousness. The latter phrase refers to the planetary articulation of place-specific modes of insertion into the globally unified production of both value and noxiousness. Since their beginning in the 1990s, the UN's Conferences of the Parties (COPs), which were established to fight climate change, have failed to secure a decline in greenhouse gas emissions. This defeat can be explained by the fact that such attempts were conducted under the project of an 'ecological transition from above', that is, a vision of ecological transition that does not question the centrality of capital's profit imperative.[4] The neoliberal, market-driven version of the ecological transition from above entered a long crisis between 2008 and 2020, a period in which China's state interventionist model proved able to narrow the dragon's technological gap with the West. The first response to this, coinciding with the Bidenomics period, can be dubbed the 'green' plan of capital. Here, a more proactive industrial policy was deployed to spur the development of renewables, electric vehicles and advanced digital technologies. Trump's return to the White House can be interpreted as a symptom of the lack of success of this policy framework. The ascendancy of the far right thus changed the plan of capital from green to white, with state intervention put at the service of a more explicit white suprematism.

In light of these transformations, the paradox of noxious deindustrialisation examined in Chapter 1 can be seen under a different light. Far from being a homogeneous phenomenon, the specific forms that noxious deindustrialisation takes at a local level depend on a whole host of contextual factors. In this respect, a key element is the mode of insertion of a given space in the international division of labour and noxiousness.

4 Lorenzo Feltrin and Emanuele Leonardi, 'Working-Class Environmentalism and Climate Justice: The Challenge of Convergence Today' [2022], *Project PPPR*, 15 February 2023, projectpppr.org; Paola Imperatore and Emanuele Leonardi, *L'era della giustizia climatica: Prospettive politiche per una transizione ecologica dal basso*, Naples-Salerno: Orthotes, 2023.

The last section of this chapter provides an example of such a specificity through a case study of the Ventanas industrial area, situated in Quintero-Puchuncaví, on the coast of central Chile. Here, the experience of noxious deindustrialisation was marked by Ventanas's historical role as a copper-processing hub within an extractivist economy, dependent on primary or 'quasi-primary' product exports.[5]

The case of Ventanas is highly significant in our times of a new scramble for natural resources among world powers. Copper is central to energy infrastructure, and its price has been on the rise even in the absence of an energy transition from fossil fuels to renewables, due to the current reality of an *energy expansion*. High copper demand boosted Chilean GDP growth, but it did not alter its subordinate position in the global socioecological hierarchy. The extractivist nature of Chile's insertion in the international division of labour has meant that noxious deindustrialisation there is characterised by relatively limited endogenous technological capabilities, a related scarcity of good-quality employment alternatives, and persistent environmental destruction. This points to the need for a truly ecological transition, one that, in addition to innovating technologies, would transform the social relations shaping technological innovation itself.

The mysteries of the Trinity formula

In *Capital*, volume 3, Marx wrote: 'Capital-profit (profit of enterprise plus interest), land-ground-rent, labour-wages, this trinity form holds in itself all the mysteries of the social production process.'[6] The trinity formula 'profit-rent-wage' thus also holds in itself the mysteries of the international division of labour and noxiousness. In Marx's frame, income is a claim on value generated by labour and is divided up into

5 Eduardo Gudynas, *Extractivisms: Politics, Economy and Ecology*, Nova Scotia: Fernwood, 2021.

6 Karl Marx, *Capital: A Critique of Political Economy*, vol. 3, London: Penguin, 1991 [1894], 254–72.

wages (and other forms of labour income treated here under the 'wage' label for the sake of brevity) returning to the workers and surplus value appropriated by the capitalists as profit and by the landlords as rent. Based on Marx's critique of the 'trinity formula', this section uncovers three main modes of insertion into the international division of labour and noxiousness: high-tech (capital-profit), low-tech with cheap labour (labour-wages), and low-tech with cheap nature (land-ground-rent).

Let us begin with profit. Tackling uneven development and imperialism, *dependentista* authors such as Ruy Mauro Marini and Enrique Dussel highlighted how the equalisation of the *profit* rate, coupled with international disparities in the development of *capital*, generated a systematic transfer of surplus value from 'peripheral' to 'core' countries.[7] As an enduring legacy of colonisation, core countries dominate the frontier of technology and thus the production of advanced capital goods (products used to make other products) such as machinery, computers and software.[8] Peripheral economies therefore usually feature lower average 'organic compositions of capital' (i.e. they are more labour intensive) or depend on technology imports from the core if they are to raise their capital compositions.[9]

According to Marx, other things being equal, the products of capital-intensive branches, for example advanced weapons, trade above their value, while the opposite is true for labour-intensive branches, such as textiles.[10] Imagine a high-tech weapons manufacturer and a textile company with similar 'variable capitals' (wage bills) but different capital intensities. Featuring the same magnitude of average living labour (averaging out, that is, skill-related wage differentials), the two firms produce

7 Enrique Dussel, *Towards an Unknown Marx: A Commentary on the Manuscripts of 1861–63*, Abingdon-on-Thames: Routledge, 2001 [1988]; Ruy M. Marini, *The Dialectics of Dependency*, New York: Monthly Review Press, 2022 [1973].

8 Theotônio Dos Santos, *Socialismo o fascismo: El nuevo carácter de la dependencia y el dilema latinoamericano*, Buenos Aires: Periferia, 1972.

9 Otto Bauer, *The Question of Nationalities and Social Democracy*, Minneapolis: University of Minnesota Press, 2000 [1907].

10 See Marx, *Capital*, vol. 3, 273–301.

the same amount of value. However, the value produced by the weapons-manufacturing workers is spread on, say, 100 units of 'constant capital' (the cost of means of production), while the equal value produced by the textile workers is spread on 50 units of constant capital only. Automated high-precision lasers are dearer than manual sewing machines. If the products of the two firms traded at their value, the textile company would enjoy a higher profit rate than the weapons manufacturer, as it would earn the same revenue while incurring less capital costs. However, as capital is always on the lookout for the highest profit rates, investment will flow from weapons into textiles, thankfully diminishing the output of killer drones and increasing that of blue jeans (unfortunately, in the real world, the margins of advanced weapons manufacturers are widened by the surplus profits guaranteed by their monopoly over cutting-edge 'defence' technology, but let us forget about this for a moment). In turn, this will raise drone prices and lower those of jeans, until the profit rate between the two branches is equalised.

There is then a tendency for the rate of profit to be equalised through price formation. Rather than selling at their value, commodities tend to sell at their price of production (costs plus average profit rate).[11] While the unevenness between our two firms is rooted in production, in their capital compositions, surplus value is transferred from the textile company to the weapons manufacturer via exchange. As Enrique Dussel put it: 'Competition . . . *does not create value*, it rather *redistributes value* starting from the *levelling of prices*.'[12] George Caffentzis summarised: 'The capitalist system demands from each capitalist according to his/her exploitation of workers [variable capital], and gives to each capitalist according to his/her investment [variable plus constant capital].'[13]

It will not escape the reader that advanced weapons are produced in the Global North more than textiles are. Geographically uneven average capital compositions are the basis of an 'unequal exchange' that, coupled

11 Ibid., 751–950.

12 Dussel, *Towards an Unknown Marx*, 347.

13 George Caffentzis, *No Blood for Oil! Essays on Energy, Class Struggle, and War (1998–2016)*, New York: Autonomedia, 2017, 29.

with profit-repatriations by capitals from the core invested in the periphery, results in the fact that capitalist development reproduces the core–periphery relationship rather than eroding it. This sort of unequal exchange can be called 'capital-based' unequal exchange. While highlighting extreme wage differentials between core and peripheral labour, Marini and Dussel saw these differentials as deriving from uneven levels of technological development, with low wages forced upon peripheral workers to compensate for 'their' lower productivity in terms of output.[14]

In contrast, moving on to the wage, both Arghiri Emmanuel (an author close to dependency theory) and the *operaistas* emphasised a dialectic between class struggle and technological development in wage formation.[15] According to Marx, the price of *labour power* (the *wage*) does not reflect labour productivity, it rather oscillates around the value of labour power, that is, the cost of its reproduction. In the determination of the latter, Marx saw a 'historical and moral element'.[16] This element is *to some extent* contingent, as shown by the fact that wage levels and wage dispersion vary across firms with similar technological endowments, and across countries with comparable development levels. The wage is *political*, not in the narrow sense of being related to state institutions, but *as an expression of power relations subject to change via social conflict*.

Emmanuel theorised 'wage-based' unequal exchange as the transfer of value from low-wage to high-wage firms (at similar levels of labour skill and intensity).[17] Imagine an aircraft manufacturer and a metal components supplier have the same capital intensity but the former pays

14 Dussel takes issue with Marini's formulation that 'the foundation of dependency is the super-exploitation of labour'. However, it seems clear from Marini's work that – just like Dussel – he considered low wages to be a consequence of laggard endogenous technological development.

15 Arghiri Emmanuel, *Unequal Exchange: A Study of the Imperialism of Trade*, New York: Monthly Review Press, 1972 [1969]; Mario Tronti, *Workers and Capital*, London: Verso, 2019 [1966]; see also Samir Amin, 'The End of a Debate' [1973], in *Imperialism and Unequal Development*, New York: Monthly Review Press, 1977, 181–252.

16 Karl Marx, *Capital: A Critique of Political Economy*, vol. 1, London: Penguin, 1976 [1867], 275.

17 Emmanuel, *Unequal Exchange*, 61–4.

its welders more than the latter. All welders produce the same amount of value per hour worked, but the metal components workers' low wages retain for them a smaller proportion of this value relative to the aircraft workers' high wages. The low-wage workers thus deliver to capital more surplus value than the high-wage ones. However, this extra surplus value is not fully captured by the metal components firm. It rather joins the global pool of surplus value to be redistributed 'equally' to total capital, as here too capital's movement in search of the highest margins tends to equalise profit rates via price formation. Investment will flow into metal components, eventually cheapening them relative to finished aeroplanes. Through this 'wage-based' unequal exchange, high-wage firms receive more labour time for less.[18] As Emmanuel summarised: 'Different rates of surplus value [rates of exploitation] combined with the same "rate of profit" means different wages together with the equalization of profits.'[19]

In an economy where workers freely compete for jobs, wage-based unequal exchange cannot last long, as labour flows to the higher-wage jobs as rapidly as possible (just like capital flows to the higher-profit firms), equalising wages for similar skills. In our example, many component-making welders would figure out they are being ripped off: it is a much better deal to quit their old low-paid job for a new and more lucrative one in the aircraft business. Unless successful class struggles equalise wages at their higher level, the aeroplane-making firms will then present their welders with a choice between taking a wage cut or being replaced with their cheaper colleagues, so that wages in the two sectors converge. However, if labour market competition is systematically hindered, wage-based unequal exchange becomes systematic too. Indeed, even if capital accumulation is globally unified, such that the freer movement of capital pushes towards a global average profit rate, labour mobility is restricted by racialised national borders.[20] Obviously, this does not imply a full

18 Jason Hickel, Dylan Sullivan and Huzaifa Zoomkawala, 'Plunder in the Post-Colonial Era: Quantifying Drain from the Global South Through Unequal Exchange, 1960–2018', *New Political Economy* 26(6), 2021, 1030–47.

19 Emmanuel, *Unequal Exchange*, 91.

20 Gloria Anzaldúa, *Borderlands/La Frontera: The New Mestiza*, San Francisco: Aunt Lute, 1987.

international immobility of labour, but, rather, a structuring of mobility that prevents global wage convergence.[21] If the low-wage welders are in Morocco and the high-paid welding jobs are in the US, then the Moroccan welders might have to accept being ripped off for a bit longer, possibly for their entire lives.

Wage differentials for similar skills also exist within states, because of the difficulties of movement from one region to another and, more significantly, because of race and gender discrimination. It is hard, for instance, to become accepted as a woman welder. The threat of sexist harassment is always there, and shift work and work-related travelling keep welders away from caring responsibilities that, more often than not, disproportionately fall upon female shoulders. Hence the widely documented gender pay gap. Nonetheless, such domestic wage divergences are smaller in scale than those among national economies.[22] Borders act as discrete fractures on the earth, breaking the unity of the total working class. The vast majority of the world's 300 million migrants will easily recognise the persistent significance of national borders, having experienced the difficulties of obtaining or renewing visas or, worse, having to get by as undocumented 'aliens'.

By way of further clarification: one hour of labour in the Global South produces as much value as one hour of labour in the Global North, at similar levels of skill and intensity.[23] It is true that countries dominating the technological frontier feature larger shares of skilled labour, which has higher reproduction costs and thus commands better wages irrespective of racialised borders. However, it is also clear that similarly skilled workers across the core–periphery divide receive vastly different wages, and this is explained by border-enforced divergences in the value of their labour power. For example, truck drivers in Peru and in the US have manifestly different real wages and living standards. The US

21 Sandro Mezzadra and Brett Neilson, *Border as Method, or, the Multiplication of Labor*, Durham, NC: Duke University Press, 2013.

22 John Smith, *Imperialism in the Twenty-First Century: The Globalization of Production, Super-Exploitation, and the Crisis of Capitalism*, New York: Monthly Review Press, 2016.

23 See Marx, *Capital*, vol. 1, 137.

teamsters' higher wages are derived neither from their higher output productivity nor from their higher skills (indeed, it might well take more dexterity to drive on Peruvian roads), but mainly from a larger, border-protected 'historical and moral element' in the value of their labour power.

Similarly to Emmanuel, the *operaistas* stressed that the wage is not only an expression of the capital–labour power relation but also a vector of power relations *within* the working class – a vector of intra-class stratification and division. This is why wage egalitarianism was an *operaista* tenet. Not merely as an ethical commitment, but as a strategic step towards working-class recomposition through the overcoming of its material divisions. The most famous expression of such a 'wage-ism' was the *operaista* strategy of the late 1960s that sought to break the Keynesian supposed proportionality between wages and productivity, with inversely proportional wage increases pushed so high as to be incompatible with the reproduction of capitalism. As Massimo Cacciari wrote: 'Essentiality of the wage struggle as antagonism to capitalist accumulation, as struggle *for the crisis* of the capitalist mode of production.'[24] This was synthetised in the slogans 'More money, less work' and 'We want everything'. And, during the expansive phase of Italy's Long 1968, wage dispersion was compressed, that is, wage inequality was actually reduced.

In his obituary of *operaista* factory leader Italo Sbrogiò, Negri recalled the debates of the time as follows:

> They were the first struggles, those of Porto Marghera's Petrolchimico, in which wage equality between blue-collar and white-collar workers – that is, among all the workers of the factory – and the refusal of union bargaining [monetisation] of noxiousness became the primary objectives of class struggle in the factory. Such demands erupted at Pirelli, Alfa Romeo, and finally in the 'Hot Autumn' at FIAT, configuring the communist aspect of a mass movement for political transformation. Italo was conscious of all this. He laughed when somebody

24 Massimo Cacciari, *Ciclo capitalistico e lotte operaie: Montedison, Pirelli, Fiat 1968*, Venice: Marsilio, 1969, 11.

> attacked him saying that these demands were economic, not political. He patiently explained that inequality is not an insult to morality, but the mode of organisation of exploitation.[25]

Initially, working-class stratification was understood mainly as it was expressed by salary grids, with pay scales assigning different remunerations based on supposed skill levels. Accordingly, the analysis remained confined to waged workers in a national context. It was therefore feminism that uncovered the gendered and racialised dimensions of wage hierarchies, showing how the working class is not only stratified by wage differentials, but also by the very presence or absence of a wage.[26] Selma James, in particular, took this analysis further by applying it on a global scale:

> The hierarchy within the working class is by no means confined to the power of men, identified with the wage, over women, identified by wageless and therefore invisible work. There is also the power of the waged worker in the metropolis over the unwaged worker in the Third World. Both are fundamental to the capitalist division of labor nationally and internationally.[27]

This interpretation of the wage as an expression of inter-class power (between the working class and capital) and intra-class power (between different working-class segments) is the background from which Ferrari Bravo read Emmanuel, leading him to write:

> The point to grasp here is the contradictoriness of the claim of continuing to construct an 'economic' theory of the price of labor-power, and the acknowledgement, which is present in Emmanuel, of the exogenous character of the 'wage' (relative to the level of development achieved by

25 Antonio Negri, 'In memoria di Italo Sbrogiò', *EuroNomade*, 1 December 2016, euronomade.info.

26 Mariarosa Dalla Costa and Selma James, *The Power of Women and the Subversion of the Community*, Bristol: Falling Wall Press, 1972.

27 Selma James, *Sex, Race and Class: The Perspective of Winning. A Selection of Writings (1952–2011)*, Oakland, CA: PM Press, 2012, 102.

> the productive forces) not merely as a biological and ethical-historical fact, but as the fact of an irreducible political subjectivity.[28]

Operaismo thus posited class antagonism as an element of contingency having the capacity to reduce intra-working-class power differentials to increase the power of the total working class versus capital. These insights remain useful today. However, it should also be noted that *operaista* 'wage-ism' sometimes lost sight of the dialectical relationship between class struggle and capitalist development, drifting towards a voluntarism in which almost anything is possible if the struggle is hard enough.

What Ferrari Bravo considered a 'two-pronged movement which unites [*accomuna*] the metropolitan worker and the proletariat of the Third World' was eventually defeated by the capitalist restructuring of the early 1970s, which, with the aid of microchips and containers, accelerated the outsourcing of large chunks of industrial production to low-wage countries.[29] In parallel, the Gulf petro-monarchies reinvested their petrodollar rents in US financial markets rather than in raising the value of labour power in their own region.[30] Workers in the Global North perceived more expensive imports as a threat to their own wages and, with the ebb of egalitarian mass mobilisations, some began to see in nationalism a better guarantor of their interests. Indeed, the divisive nature of the global wage hierarchy should be taken seriously to avoid solely cultural explanations of working-class support for the right.

The turn in the 1970s cannot be properly understood without the last element of the trinity formula, *ground rent*. Ground rent was insufficiently explored by both dependency theory and *operaismo*, as each relied on a similar labour–capital dualism. This dualism is reflected by the *dependentista* core–periphery schema, which appeared effective for analysing the international division of labour so long as 'extractivist' countries (that is, those mainly dependent on primary product exports)

28 Ferrari Bravo, 'Old and New Questions'.

29 Ibid.; Beverly J. Silver, *Forces of Labor: Workers' Movements and Globalization Since the 1870s*, Cambridge: Cambridge University Press, 2003.

30 See Midnight Notes Collective (eds), *Midnight Oil: Work, Energy, War (1973–1992)*, New York: Autonomedia, 1992.

roughly coincided with low-wage countries. However, the crises and conflicts of the 1970s brought about two key transformations: some extractivist countries became able to gain control over higher rents (e.g. the members of the Organization of the Petroleum Exporting Countries, OPEC), while East Asia became a major exporter of (low-wage) manufactured goods. These tectonic shifts mean that a dualist capital–labour/core–periphery approach is no longer sufficient. Today the 'periphery' is roughly split between areas mainly exporting rent-bearing primary products and those exporting low-wage manufactured goods. As noted by Fernando Coronil, Henri Lefebvre had already pointed out in the 1970s that the capital–labour dualism is too narrow because it pushes nature (on which ground rent is directly based) out of the analytical picture.[31]

Ground rent stems from the ownership of resource-rich *land*. Here, labour has a higher-than-average output productivity not because of cutting-edge technologies, but thanks to natural conditions not reproducible by human work.[32] Such conditions allow products to sell above their prices of production. Let us take the example of copper. Once popular for cookware, the red metal is now found in everything electric, as it has not yet been dethroned from the seat of best conductor in charge. However, the world's best copper reserves are concentrated in Chile and Peru. Were Italy to reopen the small, near-exhausted copper mines of the Alpine Dolomites that were once exploited by the Republic of Venice, then one working hour there would extract less and worse copper than one working hour in Chile's gigantic Chuquicamata mine – no matter the sophistication of the technology deployed. As the whole world demands copper, but only Chile can supply it in the required quantities, the companies extracting Chilean copper can charge a markup beyond the average profit (to an extent depending on the going market rate).

31 Fernando Coronil, *The Magical State: Nature, Money, and Modernity in Venezuela*, Chicago: University of Chicago Press, 1997, 56–66; Henri Lefebvre, *The Production of Space*, Oxford: Blackwell, 1991 [1974].

32 See Javiera Rojas Cifuentes, Gabriel Rivas Castro, Mauricio Fuentes Salvo and Juan Kornblihtt, *La cuantificación del desarrollo histórico del capital en América del Sur*, Santiago de Chile: Ariadna, 2023.

As highlighted by *iñiguismo*, resource-rich countries with laggard endogenous technological development tend to be inserted into the international division of labour as providers of rent-bearing primary commodities (predominantly scarcely-processed food, minerals or hydrocarbons).[33] Scrambles for rent appropriation generate conflicts between national and foreign actors, including the state-as-landlord.[34] When the latter manages to recover a significant share of rent, the country can escape the coercion to compete in global markets through low wages. That said, the resulting overvaluation of the national currency keeps industry uncompetitive and dependent on subsidies from rent itself.[35] To stick with our example: as of writing, Chilean wages are still higher than Chinese ones even if China has developed mightier technological capabilities. Chinese copper smelters are thus more competitive than Chilean ones, therefore most of the latter could only gain lower-than-average (and even negative) profit rates if left to their own devices. If Chile wants to keep some copper smelting within its borders, then it must subsidise the profits of its copper-processing capitals with a portion of the ground rent it generates by exporting scarcely-processed copper concentrate.

Iñiguismo has the merit of having updated Marx's theory of rent to provide original insights into today's international division of labour.[36] However, Iñigo Carrera's rejection of Emmanuel's wage-based unequal exchange seems unconvincing. Iñigo Carrera claims that national wage differentials for 'the *same product* . . . keep existing [only] as long as competition does not impose their dissolution'.[37] This is true but

33 Juan Iñigo Carrera, *La renta de la tierra: Formas, fuentes y apropiación*, Buenos Aires: Imago Mundi, 2017.

34 Nicolás Grinberg, 'Latin American Development in Historical Perspective: Capital Accumulation Through Primary-Commodity Production and Ground-Rent Appropriation', *Historical Materialism* 31(4), 2023, 45–89.

35 Franco Galdini, '"Backward" Industrialisation in Resource-Rich Countries: The Car Industry in Uzbekistan', *Competition and Change* 27, 2023, 615–34.

36 See Greig Charnock and Guido Starosta, *The New International Division of Labour: Global Transformation and Uneven Development*, London: Palgrave Macmillan, 2016.

37 Iñigo Carrera, *La renta de la tierra*, 218, emphasis added.

misses the point, because Emmanuel specifies that unequal exchange is maintained by international trade among *different products*. In fact, according to Emmanuel, international competition to sell the same product 'is the most powerful check on an unlimited increase in wages in the advanced countries, but it is also the most powerful check preventing a take-off of wages in poor countries'.[38] However, as Emmanuel goes on, 'a high-wage country can never find itself in a position where it cannot discover a specialization that, in the international division of labor at that moment, is free from competition on the part of the low-wage countries'.[39] Artificial intelligence is just the latest item in a long list that began, perhaps, with industrially woven cotton. Furthermore, Iñigo Carrera's denial of international wage differentials for similarly skilled labour flies in the face of the empirical evidence.[40]

Bringing all this together: *quantitatively*, the value outflows generated by laggard technological development or low wages (what I have called 'capital-based' and 'wage-based' unequal exchange) may or may not be mitigated or even exceeded by rent inflows. Unequal exchange, profit repatriations and rent thus generate an intricate, multidirectional web of value transfers, whose net flows are difficult to pin down.[41] *Qualitatively*, nonetheless, rent as a line of income is very different from profits and wages. In fact, by acting as an obstacle to industrialisation, it tends to engender high inequality and bar advanced endogenous technological development.

The 'trinity formula' criticised by Marx indicates three broad patterns of insertion in today's global economy. High-tech zones – roughly analogous with the so-called 'Global North', which is perhaps gradually being joined by China – mainly claim their share of value through capital, via profits. Their cumulative technological

38 Emmanuel, *Unequal Exchange*, 135.

39 Ibid., 145.

40 Iñigo Carrera, *La renta de la tierra*, 218; Smith, *Imperialism in the Twenty-First Century*.

41 See Jaime Osorio, 'Ley del valor, intercambio desigual, renta de la tierra y dependencia', *Argumentos* 30(83), 2017, 219–24.

advantage allows them to constantly recreate export sectors combining average or beyond-average profit rates with relatively high wages. Low-tech zones, analogous with the so-called 'Global South', must compete through the abundance of their natural resources or the misery of their wages – both compensatory mechanisms for pushing laggard capitals closer to (or even beyond) the average profit rate. The areas endowed with rich soils but deprived of advanced endogenous technological capabilities, including several countries in South America, Africa and the Middle East, mainly make their claim to value through 'cheap nature', via rents.[42] The countries whose large workforces are more decisive than their resource endowments, for example much of East and South Asia, chiefly compete through 'cheap labour', via low wages. When we look beyond the mystifying veil of the trinity formula as Marx did, however, we can see that these are all claims on the total value produced by global labour under capitalist social forms.

Of course, the whole world is crossed by class relations cutting horizontally through its population. The picture just sketched is then a highly simplistic one, as all countries, as well as sub-national and super-national zones, combine the three modes of insertion in the global economy in different and, to some extent, contingent amalgams. Nonetheless, this outline provides a frame for detailed investigation into concrete local insertions in global capital accumulation, and the related compositions of specific segments of the planetary working class.

In addition, the international division of labour is also an international division of noxiousness. This phrase echoes Fernando Coronil's brilliant analysis of the 'international division of labour and nature'.[43] However, Coronil's wording refers to labour and nature as the transhistorical producers of wealth. My terminology, instead, emphasises capitalist labour as the producer of value. As argued in the introduction,

42 See Jason W. Moore, *Capitalism in the Web of Life: Ecology and the Accumulation of Capital*, London: Verso, 2015.

43 Coronil, *The Magical State*.

value is an inherently noxious social form as it necessarily deepens the 'metabolic rift' between society and nature.

The Global North can afford stricter health and environmental regulations relative to the Global South. In the latter, cheap-nature economies rely on large-scale resource depletion, becoming 'nature-exporting societies', while cheap-labour ones are characterised by unhealthy and polluting factories.[44] Loose regulations and austere welfare services are instrumental in ensuring this cheapness, which comes at the cost of high emissions, water exhaustion, soil erosion, waste accumulation, biodiversity loss, etc., along with the dispossession of indigenous populations from ancestral commons.[45] Furthermore, as just noted, industrial production in extractivist countries tends to be subsidised by ground rent, which is spent to import outdated machineries from the Global North, lengthening their productive life. This has its own ecological impacts, because such 'scrap' fixed capital is usually more noxious than the best available technologies. The labour hierarchy and the noxiousness hierarchy are thus mutually reinforcing. In the inverted world of capital, low technological capabilities constitute a pressure towards high pollution.

A parallel can be made between domestic and international socioecological hierarchies. Within countries, the workers with low employable skills are more likely to reside close to highly polluting facilities (where housing is usually more affordable) and to work in dangerous jobs, which often provide an implicit 'noxiousness bonus'. This phenomenon is known as the 'monetisation of health'.[46] Internationally, countries with low endogenous technological capabilities are more likely to specialise in exports characterised by high noxiousness at the source, in what could be described as an international monetisation of health and nature. Similarly to wage differentials, international environmental inequalities

44 On 'nature-exporting societies', see ibid.

45 Astrid Ulloa and Barbara Göbel (eds), *Extractivismo minero en Colombia y América Latina*, Bogotá: Universidad Nacional de Colombia, 2014.

46 Lorenzo Feltrin and Devi Sacchetto, 'The Work-Technology Nexus and Working-Class Environmentalism: Workerism versus Capitalist Noxiousness in Italy's Long 1968', *Theory and Society* 50(5), 2021, 815–35.

are compounded by militarised borders restricting labour mobility (as seen in the previous chapter).

The relationship between social and ecological inequalities is evident in World Bank average life-expectancy figures: eighty years in high-income countries versus sixty-two in low-income ones (bearing in mind that national averages conceal intersectional class, racial and spatial inequalities within countries). Depending on their political histories, however, extractivist countries can allocate a share of their rents to the import of relatively advanced technologies and the improvement of regulations and services. For example, life expectancy in Chile is remarkably higher than in the US (seventy-nine versus seventy-six years in 2021), despite the obvious disparity in their incomes per capita and endogenous technological capabilities.

The present division of labour and noxiousness can be visualised by returning to the example of smartphone production. The minerals needed for the batteries, such as copper and lithium, are mostly located in South America, while cobalt is extracted in Congo, sometimes in conditions of modern slavery.[47] The hydrocarbons necessary for the plastics are disproportionately pumped out of the Middle East.[48] The most advanced models and their microchips, meanwhile, are designed in the US. EUV lithographic machines are made in Holland and shipped to Taiwan, where the vast majority of advanced microchips are 'printed', and the smartphones are then assembled mainly in China and other relatively low-wage Asian countries.[49] While noxiousness is released at every stage of the process, it is the fenceline communities and workers who come into contact with the extractive sites and the least regulated factories that bear the brunt of it.

Nonetheless, the international division of labour and noxiousness never stands still. It is constantly changing in unpredictable ways as

47 Christian Fuchs, *Digital Labour and Karl Marx*, Abingdon-on-Thames: Routledge, 2015.

48 Alice Mah, *Petrochemical Planet: Multiscalar Battles of Industrial Transformation*, Durham, NC: Duke University Press, 2023.

49 Jenny Chan, Mark Selden and Pun Ngai, *Dying for an iPhone: Apple, Foxconn, and the Lives of China's Workers*, London: Pluto Press, 2020.

capitalism adapts to the pressures of class struggle and competition. The next section examines how the so-called 'ecological transition' entered the battleground of recent reconfigurations of global socioecological hierarchies.

The plan of capital today, from 'green' to white

The most compelling critique of the ecological transition from above is simply the historical record of greenhouse gas emissions: more CO_2 has been emitted since the establishment of the United Nations Framework Convention on Climate Change in 1992 than in the two centuries before that, with no inversion of the tendency yet.[50] This reminds us that the ecological transition from above cannot be considered an ongoing process, because, unfortunately, there is no ecological transition of any kind happening. The ecological transition from above is thus better seen as a discourse that functions to prevent an ecological transition, hinging upon the promise of decarbonising the economy through capitalist technological innovation. It served as a legitimating narrative for the fourth capitalist industrial revolution.

The crisis of the ecological transition from above opened the way for a broader radicalisation of environmentalism. The year 2019 was memorable in this respect. From Chile to Sudan to France, millions of people took to the streets. While these protests contained all the ambiguities that necessarily characterise mass mobilisations, their demands constituted a challenge to both liberal commonsense and right-wing exclusionary politics. The climate strikes added an explicitly ecological movement to implicitly environmental struggles over access to water, land and health. The year 2020, however, was a brutal counterpoint: empty squares, full hospitals and working-class decomposition over the 'correct' response to the COVID-19 pandemic. If Industry 4.0 had failed to become widely diffused over the 2010s, the economic downturn, marked by the pandemic

50 Feltrin and Leonardi, 'Working-Class Environmentalism'; Imperatore and Leonardi, *L'era della giustizia climatica*.

and the increasing visibility of climate change, became – like all capitalist crises – an occasion for attempts to relaunch capitalist development on new technological bases, particularly renewable energy, electromobility and advanced digital technologies.

Between 1971 and 2021, the energy provided by renewables and waste (including biomass and hydro, as well as solar and wind) has grown more than threefold, from 29.6 to 90.8 zettajoules (see Figure 3.1). However, the increase of their share in the world's total energy supply over the same period has been underwhelming: from 12.9 to 14.7 per cent. The relative decline of fossil fuels in the global energy matrix during this half century, from 86.6 to 80.3 per cent, was due more to the rise of nuclear (from 0.5 to 5 per cent) than to renewables. So far, we are not seeing an energy transition but an energy expansion, in which fossil fuels – instead of being phased out – keep increasing at unsustainable levels, with renewables growing alongside them.[51]

Figure 3.1 World energy supply by source (zettajoules)

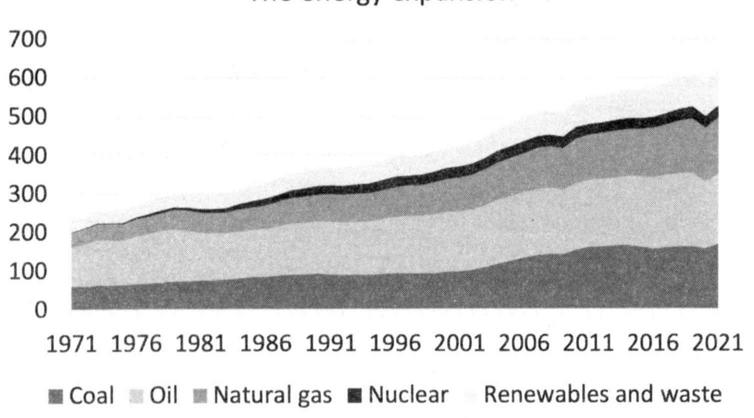

Source: IEA.

51 Sean Sweeney, John Treat and Daniel Chavez, *Energy Transition or Energy Expansion?*, New York: Transnational Institute, 2021.

Part of the problem is that energy is mostly produced as a commodity. At current inequality levels, and within the framework of market-mediated consumption in which we live, decarbonisation needs massive investment efforts that private companies are only willing to undertake if they can expect to make a normal profit out of it. The energy transition has therefore been affected by the slow growth and sputtering investment rates that currently characterise the global economy. For example, 'over-capacity' in solar panel production would be a blessing for the climate but not for capital.[52] In fact, if solar panel prices decline so quickly as to push profits below the normal rate, investment in solar panels will slow down as capital seeks more profitable outlets.

The profit imperative affects not only the quantity of renewable energy available but also its qualitative aspects. We can see it, for instance, in the development of solar megaprojects in North Africa, where sun is abundant, for export to Europe, where money is to be found.[53] In the first place, this replicates the classic international division of labour shaped by coloniality, whereby Global South countries provide the West with the raw materials it needs to keep dominating the technological frontier. However, it also means that renewables-linked research efforts are channelled into the cheapest ways to transport electricity over very long distances, particularly through marine cables called interconnectors, rather than prioritising local energy poverty alleviation. A glaring example is the Xlinks Power Project, which aims to produce renewable energy in Morocco, and possibly in disputed Western Sahara, to export it to the UK via a 3,800 km long interconnector. The distributional absurdity of this megaproject is readily shown by the fact that average per capita energy consumption in Morocco is 6,855 kilowatt-hours (even less in most nearby African countries), while it is 30,098 in uninsulated Britain. It is no wonder

52 Jack Copley, 'Decarbonizing the Downturn: Addressing Climate Change in an Age of Stagnation', *Competition and Change* 27(3–4), 2023, 429–48.

53 Hamza Hamouchene and Katie Sandwell, *Dismantling Green Colonialism: Energy and Climate Justice in the Arab Region*, London: Pluto Press, 2023.

then that experts are asking whether there will be enough copper cables for the energy transition.[54]

Moving on to electromobility, the projected rise of renewables is supposed to decarbonise transport. The diffusion of electric vehicles has propelled deep restructurings of the global automotive workforce in the name of the ecological transition from above. However, private electromobility comes with its own ecological problems: the production of electric vehicles is less labour-intensive but more energy-intensive than that of petrol cars, rechargeable batteries are difficult to recycle and an immense amount of minerals will have to be dug out of the earth.[55] Furthermore, the total number of vehicles around the world is expected to keep growing over the coming decades. Similar to the trend of renewables co-existing with fossil fuels rather than replacing them, the most likely outcome, as things stand, is an expansion of the global vehicle fleet whereby the share of combustion engines decreases, while their absolute number does not. In line with the general tendency, increased efficiency is offset by rising material output. Instead of maintaining a transport infrastructure that prioritises private cars, what is needed is a multimodal transport network centred around public mobility.[56]

American industry leader Tesla is the perfect illustration of everything that is wrong with capitalist electromobility. The promise of decarbonisation has contributed to make right-wing provocateur Elon Musk the richest man in the world. However, luxury private cars stuffed with autochips will not do much to alleviate the ecological crisis, nor to achieve a more egalitarian access to mobility. As German campaigners noted: 'Tesla boss Musk opposes public transport (long- and short-distance), cycling and pedestrian mobility just as intensely as he

54 Rachel Millard, 'Will There Be Enough Cables for the Clean Energy Transition?', *Financial Times*, 29 July 2023.

55 Jörn Boewe and Johannes Schulten, *The Transformation of the Global Automotive Industry: Trends, Interpretations, Socio-Ecological Strategies for Action*, Châtelaine: Rosa-Luxemburg-Stiftung, 2023.

56 Ulrich Brand and Markus Wissen, *The Imperial Mode of Living: Everyday Life and the Ecological Crisis of Capitalism*, London: Verso, 2021 [2017], 135–59.

counters the unionisation of his factories.'[57] The socioecological hierarchies implicit in Tesla's vision for mobility – expensive e-cars for the affluent, union rights suppression for the workforce and pollution for mining fenceline communities – are notoriously made clear by Musk himself.

E-cars are also one of the many applications of advanced digital technologies. Electric vehicles are often described as 'computers on wheels', to be driven one day by artificial intelligence. Machines of all kinds (such as industrial robots, household and office appliances or military equipment) are linked through the internet of things and 5G connectivity to software applications (artificial intelligence, big data analytics, blockchains, etc.), generating networks of 'computerised things'. Such digitalisation has generated the illusion of a 'dematerialisation' of production, which would generalise 'smart' jobs that offer easier lives for workers while being friendly to the environment.[58] The reality, however, is that digitalisation has facilitated the rise of precarious jobs in the gig economy, while the computerisation of everything needs raw materials and energy, such that 'smart' jobs in one place are still dependent on 'dirty' ones elsewhere.[59]

Amazon, for example, has promoted the state-of-the-art digitalisation of its warehouses through a culture of happiness at work and green techno-fixes.[60] The Seattle giant's technocratic approach does reduce some aspects of internal and external noxiousness. Regarding internal noxiousness, robots move heavy objects around, sparing workers' backs from the burden. In terms of external noxiousness,

57 Klima AG Berlin, 'Tesla Gigafactory: Ein Elektromotor macht noch keine Verkehrswende!', interventionistische-linke.org., 21 April 2020.

58 Ursula Huws, *Labor in the Global Digital Economy: The Cybertariat Comes of Age*, New York: Monthly Review Press, 2014.

59 Marco Marrone, *Rights Against the Machines! Il lavoro digitale e le lotte dei rider*, Milan: Mimesis, 2021; Jamie Woodcock, *The Fight Against Platform Capitalism: An Inquiry into the Global Struggles of the Gig Economy*, London: University of Westminster Press, 2021; Giorgio Pirina, *Connessioni globali: Una ricerca sul lavoro nel capitalismo delle piattaforme*, Milan: FrancoAngeli, 2022.

60 Alessandro Delfanti, *The Warehouse: Workers and Robots at Amazon*, London: Pluto Press, 2021.

Amazon projects a 'green' image through the installation of solar panels on its warehouses and data centres, and the promotion of electric vehicles among its outsourced drivers. However, both sides of noxiousness, internal and external, come back with a vengeance. Internally, the strict surveillance enabled by tracking devices minimises 'wasteful' ebbs in the working day, thereby maximising labour intensity.[61] This causes a conspicuous incidence of injuries and burnout among Amazon workers, as well as very high turnover rates.[62] Amazon responded to these challenges with admirable coherence. Rather than dropping its anti-union policies, it rolled out AmaZen, a wellness booth (or 'despair closet', in the words of its critics) equipped with vegetation and a computer that workers can use to meditate their exploitation away.[63] Externally, Amazon's general business model is predicated on the intensification of the interconnected productivist and consumerist tendencies inherent to capitalism. The overall result is an utterly unsustainable carbon footprint.[64]

In sum, the dialectic between technological development and social relations holds for renewable energy, electromobility and advanced digitalisation too. In the abstract, all of these technologies – quantitatively contained and qualitatively transformed – would be useful in an eco-socialist society. However, the concrete ways in which they are being developed results from the clash between class struggle and the prevailing commodification of production and nature. As set out in the introduction, the law of value shapes the *how much*, the *how* and the *what* of production. The resulting inequalities include an unbalanced international division of labour, in which, as the previous section described, some areas participate by disproportionately contributing

61 Sarrah Kassem, *Work and Alienation in the Platform Economy: Amazon and the Power of Organization*, Bristol: Bristol University Press, 2023.

62 Beth Gutelius and Sanjay Pinto, *Pain Points: Data on Work Intensity, Monitoring, and Health at Amazon Warehouses*, Chicago: University of Illinois, 2023.

63 Delfanti, *The Warehouse*, 59.

64 Will Evans, 'Private Report Shows How Amazon Drastically Undercounts Its Carbon Footprint', *Reveal*, 25 February 2022, revealnews.org.

cheap nature, others cheap labour, and others cutting-edge technology in a steeply uneven mix.

The competition between the US and China for the control of the technological frontier has exacerbated the fight over 'critical minerals' (or 'critical raw materials'), such as cobalt, copper, lithium and nickel. With some exceptions, their extraction is currently concentrated in certain areas of Latin America, Africa and Asia (see Figure 3.2). If the forecasts made by countless reports and policy briefs turn out to be accurate, then the mismatch between critical mineral demand and supply will widen greatly, boosting ground rent, altering its geographical flows and exacerbating conflicts over its appropriation. The unequal relationship between the high-tech countries competing for raw materials and the low-tech countries in which they are situated is even sometimes betrayed by the rhetoric of mainstream commentators. A *Financial Times* editorial, for example, insisted that 'the west needs to bolster its efforts in the global *scramble* to secure metals and minerals that are critical for the green transition and new technologies'.[65] It is difficult not to be reminded of the late nineteenth-century Scramble for Africa, in which European powers competed to colonise Africa and take over its natural resources to develop industry at home.

The concept of 'critical mineral' has its roots in the First World War, when the warring powers began drafting lists of 'war materials' deemed essential for waging a fully industrialised conflict.[66] In that context, critical minerals were those considered necessary for national security (a notion far from neutral) but whose supply chains were easily subject to disruption. Because of this security–fragility nexus, various powers believed that acquiring critical minerals could not be left to ordinary

65 'Three Inconvenient Truths About the Critical Minerals Race', *Financial Times*, 14 July 2023, emphasis added.

66 Roy MacLeod, 'The Mineral Sanction: The Great War and the Strategic Role of Natural Resources', in Richard P. Tucker, Tait Keller, John R. McNeill and Martin Schmid (eds), *Environmental Histories of the First World War*, Cambridge: Cambridge University Press, 2018, 99–116.

Figure 3.2 Top extracting countries for selected critical minerals, % of global production

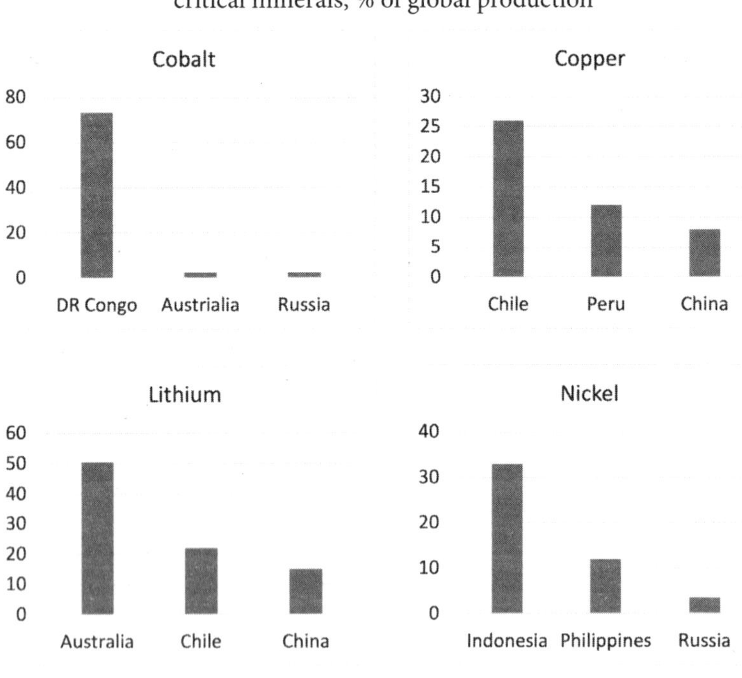

Source: IRENA 2022.

market mechanisms, but instead required recourse to 'extra-economic' measures, ranging from diplomatic activism to military invasion. This *politicisation of nature* can still be regarded as a distinctive feature of critical minerals. In the 2010s, however, a third dimension gained prominence alongside security and fragility: 'criticality' for the ecological transition. Yet Russia's invasion of Ukraine and, more generally, the diffusion of armed conflicts in the early 2020s restored the centrality of 'war preparedness'. With the shift away from the green agenda and towards rearmament, the original definition of critical minerals has again become the most relevant. Indeed, lithium, cobalt, copper, nickel and rare earths – just to name the most well-known examples – are needed not only for renewables but also for advanced digital technologies and cutting-edge weapons systems.[67]

Technological innovation drives tectonic shifts in extractive frontiers. Eduardo Gudynas coined the phrase 'new extractivism', or 'neo-extractivism', to describe the policy regime that became widespread across Latin American countries with left-wing governments.[68] While in 'classic extractivism' ground rent was chiefly appropriated by Western multinationals and their partners within the local elites, in neo-extractivism the state steps up its rent share – through nationalisations, mining royalties, export taxes, etc. – to finance redistributive policies.[69] A conservative variety of neo-extractivism is to be found in the bastions of fossil capital, the Gulf petro-monarchies.[70] Here too, the state has a large role in rent appropriation, through land ownership and SOEs, with Saudi Aramco famously being one of the most profitable companies in the world. However, its redistributive function is much more limited, as most of the Gulf workforce is comprised of super-exploited immigrants, whose labour conditions often border on slavery. A real energy transition would minimise demand for oil and gas, and thus annihilate the 'grey' carbon rents reaped by the petro-monarchies. Unsurprisingly, the latter have been staunch allies of the Western climate-denialist right.

In any case, ground rent, whether 'green' or 'grey', is a mixed blessing. Mining has notoriously noxious effects on fenceline communities, and the 'green' mining required to supply the ecological transition from

67 Sophia Kalantzakos (ed.), *Critical Minerals, the Climate Crisis and the Tech Imperium*, New York: Springer, 2023.

68 Eduardo Gudynas, 'The New Extractivism of the 21st Century: Ten Urgent Theses About Extractivism in Relation to Current South American Progressivism', *Americas Program Report*, 2010, 1–14.

69 Maristella Svampa, *Neo-Extractivism in Latin America: Socio-Environmental Conflicts, the Territorial Turn, and New Political Narratives*, Cambridge: Cambridge University Press, 2019.

70 Adam Hanieh, *Money, Markets, and Monarchies: The Gulf Cooperation Council and the Political Economy of the Contemporary Middle East*, Cambridge: Cambridge University Press, 2018.

above is no exception.[71] What is more, fenceline communities rarely reap much benefit from it. The core mining workforce tends to be skilled and highly deterritorialised, so that the locals mostly get jobs in the low-paid services linked to mining or continue working in sectors independent from it.[72] Furthermore, extractivist economies are excluded from advanced technological development, because the appreciation of their currencies prevents them from further processing their raw materials competitively. As a result, the productivity gap between such countries and the Global North is far from closing.[73]

Contemporary extractivism thus reproduces in new ways the patterns of the classic international division of labour; and yet, in terms of the industrial processing of these raw materials, things are moving in a different direction. Since the neoliberal turn of the 1970s, a new international division of labour has emerged, in which East Asia and China in particular have become the workshop of the world. Even if China is not the top global miner of many critical raw materials, it has a key role in refining them (see Figure 3.3) and turning them into intermediate and final products, such as solar panels, rechargeable batteries and electric vehicles. As seen in the recent clashes between the Trump and Xi Jinping administrations, China has an even stronger grip on rare earths, necessary for a wide variety of high-tech components.

China's advantage on 'green' tech has disquieted Western policymakers. The trade wars waged by both Trump and Biden are also geared at preventing the Asian giant from accessing and ultimately developing cutting-edge technologies. However, claims that China worryingly 'dominates' the 'green' transition betray a number of problematic

71 Martín Arboleda, *Planetary Mine: Territories of Extraction Under Late Capitalism*, London: Verso, 2020; Miriam Lang, Breno Bringel and Mary Ann Manahan, *Más allá del colonialismo verde: Justicia global y geopolítica de las transiciones ecosociales*, Buenos Aires: CLACSO, 2023.

72 Gerardo Hernández Román and Jorge Pavez Ojeda, 'Neoliberalización y flexibilidad en el mundo del trabajo: Notas sobre los trabajadores de la minería en Chile', *Sociedad hoy* 23, 2012, 49–66.

73 International Labour Organization, *World Employment and Social Outlook: Trends 2023*, Geneva: International Labour Office, 2023, 90–2.

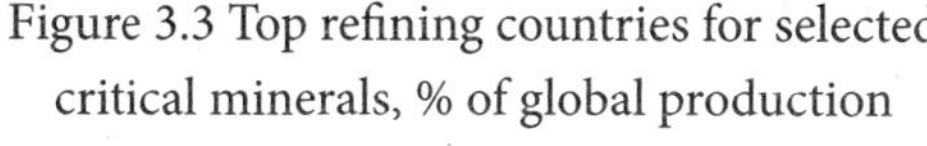

Figure 3.3 Top refining countries for selected critical minerals, % of global production

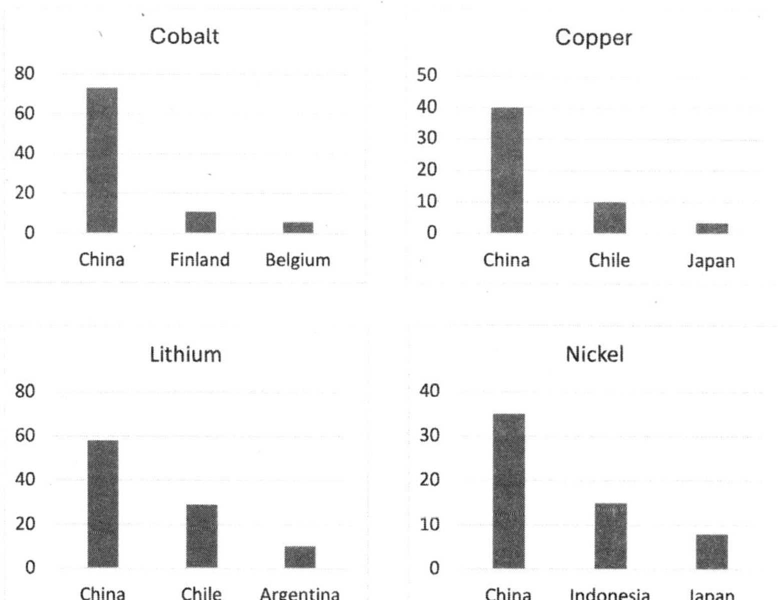

Source: IRENA 2022.

assumptions.[74] First, as argued throughout this book, as long as transition attempts remain within capitalist parameters, they are unlikely to deliver an adequate response to the ecological crisis. Second, it is actually the Global North that still dominates the technological frontier and thus the global economy. For example, Western companies are the ones able to design the most avant-garde microchips (e.g. Silicon Valley's Nvidia) and produce the latest generation EUV lithographic machines required to print them (Holland's ASML). This is why China's Huawei received a real blow when the first Trump administration cut off its access to these technologies. As things currently stand, China's large share of 'clean' tech industrial production still relies on a combination of cheap labour and

74 See, e.g., Edward White, 'How China Cornered the Market for Clean Tech', *Financial Times*, 8 August 2023.

cheap nature, in addition to rising technological prowess, making it more competitive than the Global North.[75]

Nevertheless, China has been narrowing the technological gap separating it from the Global North, and might therefore be on a trajectory towards 'dominance'. To stick with the former example, Huawei surprised Western policymakers and business leaders when it proved able to produce relatively advanced microchips even with outdated deep ultraviolet (DUV) machines.[76] However, and this is the third objection to the mainstream narrative, it is not clear why a division of labour dominated by the West should be seen as inherently better. After all, China's GDP per capita is still much lower than that of Global North countries, such that its catchup is contributing to reducing global inequality. Anticapitalists are advised not to be uncritical of the People's Republic, but any claims to Western moral superiority can easily be dismissed by observing the long series of colonial wars and hierarchies imposed by European and North American powers on the rest of the world since the fifteenth century.

The neoliberal ecological transition from above, as envisioned by the COPs since the 1990s, underwent a long crisis between the 2008 financial crash and the COVID-19 pandemic. The recovery from the latter was thus marked by an important development: a relative return of direct state involvement in the economy, indicating a discontinuity from the previous neoliberal policy framework. This has been described as 'state capitalism', meaning configurations of capitalism where the state plays a particularly strong role as promoter, supervisor and owner of capital through an expansion of sovereign wealth funds, policy and development banks, SOEs, industrial policy and economic nationalism.[77] Again, China has been central here, as its state-led policy regime – to the great scorn of neoliberal ideologues – has proved to be more effective in promoting economic growth than the Western 'free market' approach. The West

75 Hickel, Sullivan and Zoomkawala, 'Plunder in the Post-Colonial Era'.

76 Qianer Liu, 'How Huawei Surprised the US with a Cutting-Edge Chip Made in China', *Financial Times*, 29 November 2023.

77 Ilias Alami and Adam D. Dixon, *The Spectre of State Capitalism*, Oxford: Oxford University Press, 2024.

eventually followed China's statist example, as seen with Biden's Inflation Reduction Act (IRA) and, to a lesser extent, the EU's Green Deal.

The *operaistas* analysed the post-1929 rise of the 'planning state' as a capitalist response to the revolutionary threat posed by the labour movement of the time. In this sense, Keynesianism represented the incorporation by bourgeois economics of 'the working class's rupture as the motive of capitalist economic growth itself'.[78] The New Deal was a capitalist internalisation of the contradiction between relations of production and workplace struggles, or what James O'Connor called the first limit of capital.[79] As argued by Emanuele Leonardi, recent 'green' growth policies can be seen as an attempt to internalise the contradiction between relations of production and struggles at the point of reproduction, that is, O'Connor's second limit of capital.[80] In both cases, state intervention is geared towards mediating and incorporating the dynamism constituted by the working class as the 'moving mover of capital', to relaunch the production of commodities on an ever-expanding scale. Yet there are limits, social and material, to the internalisation of limits. Green Keynesianism framed decarbonisation as an opportunity for growth, but relative stagnation continued, together with the unsustainable depletion of nature.

Panzieri warned that capitalist state planning, just like capitalist technology, is moulded by the profit imperative. He thus criticised the notion, popular within the left, that state planning inherently points towards an overcoming of capitalism:

> Marx's own analysis of the factory and direct production in capitalism is rich enough to offer components for the formulation of a

78 Antonio Negri, 'Keynes and the Capitalist Theory of the State Post-1929' [1968], in *Revolution Retrieved: Writings on Marx, Keynes, Capitalist Crisis and New Social Subjects (1967–83)*, ed. John Merrington, London: Red Notes, 1988, 5–42, 9.

79 Luciano Ferrari Bravo, 'The New Deal and the New Order of Capitalist Institutions' [1972], *Viewpoint Magazine*, 2 October 2014, viewpointmag.com; James O'Connor, 'Capitalism, Nature, Socialism: A Theoretical Introduction', *Capitalism, Nature, Socialism* 1, 1986, 11–38.

80 Emanuele Leonardi, *Labour Nature Value: André Gorz Between Marxism and Degrowth*, London: Verso, 2026 [2017].

> socialist perspective that does not rest on the illusory and mystified basis of its identity with planning taken in itself, i.e. abstracted from the social relationships that may find expression in its various forms. In his analysis, Marx destroys the basis for any misunderstanding regarding capitalism's incapacity for planning. He shows that, on the contrary, the system tends to react to any contradiction or limitation on its own maintenance and development by increasing its degree of planning.[81]

> Industry re-integrates in itself finance capital, and then [through state intervention] projects to the social level the form specifically assumed by the extortion of surplus value: as 'neutral' development of the productive forces, as rationality, as plan.[82]

Over the past few years, we saw something similar. The 2019–20 struggle–crisis sequence created the political space for a 'green' plan of capital in the West. This is the policy regime that emerged between 2020 and 2024, with Bidenomics in the US and the implementation of the Green Deal in the EU. It was not merely greenwashing, but the deployment of state planning to relaunch accumulation on new technological bases, a statist version of the ecological transition from above. In some respects, this was a win relative to fossil capital's neoliberal business as usual. However, these reforms were insufficient as they did not go beyond a renewed incorporation of class struggle to push forward capitalist restructuring. As Tronti wrote: 'The "plan" of capital emerges primarily from the need to make the working class function *as such* within social capital.'[83] The state-sponsored development of 'clean' tech – even when public ownership was deployed – remained largely constrained within the commodity form and thus moulded by the impersonal law of value. Despite its green intentions,

81 Raniero Panzieri, 'Surplus Value and Planning: Notes on the Reading of *Capital*' [1964], in *The Labour Process and Class Strategies* (CSE Pamphlet no. 1), London: Stage 1, 1976, 4–25, 19.

82 Ibid., 21.

83 Tronti, *Workers and Capital*, 92.

this state intervention remained *of capital* and was therefore incapable of achieving a consciously planned transition to sustainable production. This was the case even as this more statist form of the ecological transition from above presented itself, to echo Panzieri, as neutral development, as rationality, as plan.

The 'green' plan of capital was inadequate on its own terms: it not only failed to contain the ecological crisis but was also unable to make Western 'clean' tech competitive with Chinese industry. Trump's return to the White House and the European rearmament plan are symptoms of this defeat, and they mark the abandonment of the fading green façade of the new state interventionism. However, we are not returning to the neoliberal *status quo ante* either. Policies such as the weaponisation of tariffs, the increase of state ownership in strategic industries such as rare earth mining and chip design, or government activism for the sourcing of critical minerals all signal a departure from the previous policy regime. If the 'green' plan of capital marked the 2020–24 period, the conjuncture currently characterising the West can be designated the 'white plan of capital', where 'white' stands for an ever more open white supremacy. Here, state intervention no longer claims to serve the ecological transition, but explicitly aims to defend the privileged position of the Global North in the international division of labour, a position threatened by China's rise. The scramble for raw materials is not so much driven by decarbonisation as it is rushed by an increasingly explicit arms race. The West is thus attempting to avoid or slow the decline of its global hegemony with a hardening of racism inside and outside, at extremely high human cost. The genocide Israel is committing in Gaza with Western complicity is emblematic of this.

Ventanas, Chile: Noxious deindustrialisation in extractivism

On 11 September 1973, Chilean armed forces traversed the Quintero Bay and stormed the Ventanas copper smelter and refinery, breaching its fences from the beach and swiftly occupying the premises. The

CIA-backed coup d'état against Salvador Allende's democratically elected government is usually evoked by traumatic images of Hawker Hunter jets bombing the presidential palace La Moneda and soldiers dragging thousands of political prisoners to Santiago's National Stadium, where many faced torture and assassination. On the same day, however, the army also took over the country's key industries with the strategic aim of breaking the strongholds of labour militancy that had pushed towards the radicalisation of the governmental programme for a transition to socialism.[84] Half a century later, copper is centre stage again, with the increasing worry that its supply will not match demand. Socioecological conflicts around copper processing in Ventanas provide a window both into how noxious deindustrialisation plays out in an extractivist context, as well as into the historical and current transformations of the international division of labour and noxiousness.

The decision to establish the industrial complex in Ventanas, an agricultural and fishing community in the town of Puchuncaví, which borders the town of Quintero, was finalised on 31 December 1957. As part of an import-substitution industrialisation policy regime, the Chilean government intended to increase the amount of copper processed within the country. This attempt to keep a larger share of rent within Chilean borders through the creation of state-owned industries required, however, the import of outdated machinery from the Global North. The authorities thus tasked a West German consortium with building the complex and established the state-owned Empresa Nacional de Minería (National Mining Enterprise, ENAMI) to take it over. Construction began in 1960, but the smelter was only inaugurated on 30 September 1964 (after the opening of a coal-fired power station in the same year), while the refinery opened on 11 November 1966. The project did not feature an emission capture system even though the

84 Franck Gaudichaud, *Chili 1970–1973: Mille jours qui ébranlèrent le monde*, Rennes: Presses Universitaires de Rennes, 2013; Peter Winn, *Weavers of Revolution: The Yarur Workers and Chile's Road to Socialism*, Oxford: Oxford University Press, 1986.

technology already existed, and it was ridden with delays, cost overruns and accidents.[85]

Following an interviewee's terminology, the first Ventanas workers can be roughly categorised into three main groups: *nortinos* (Northerners), *pitucos* (posh) and *huasos* (peasants). The *nortinos* were mostly skilled workers from the desertic Copiapó mining region, where ENAMI already operated the Paipote smelter. Often the children of miners, they had grown up in a culture steeped in familiarity with the red metal. The legend goes that some Northern *horneros* (furnace operators) could bare-handedly assess the quality of copper samples before the analysts even set their devices on them. The Northern mines were also the cradle of the Chilean labour movement, and many *nortinos* thus brought with them a culture of working-class radicalism that found a fertile ground in the late 1960s.

The *pitucos* were formally educated technicians, white-collars and managers. Most commuted on company buses from the urban centres of Valparaíso, Viña del Mar and Quilpué. Almost all women employed by ENAMI Ventanas at the time, although a small minority of the workforce, fell under this category. However, their white-collar status afforded them little protection from sexism, so much so that the workers' journal *La unidad* (directed by shop steward Carlos García) had to publish several appeals to stop gendered harassment. As the Ventanas complex started operating under Eduardo Frei's Christian Democratic government, many *pitucos* were initially recruited from the ranks of the party membership. However, when the leftist coalition Unidad Popular (Popular Unity, UP) brought Allende to the presidency by winning the 4 September 1970 elections, a new cohort of young, left-leaning graduates and students began crossing the factory gates.

The *huasos* mostly hailed from the nearby agricultural communities. While they were the 'unskilled' labourers, with little formal education, the other employees would hardly have been able to perform the heavy

85 Mauricio Folchi, 'Historia ambiental de las labores de beneficios en la minería del cobre en Chile, siglos XIX y XX', PhD thesis, Universitat Autònoma de Barcelona, 2006, 461–78.

duties assigned to the *huasos*. With acid rains devastating fenceline farming, against which protests were held also at the time, these workers from the fields had been dispossessed by noxiousness.[86] Such dispossession by accumulation compelled many farmers to trade their sickle for a hammer in the factory. It was, in a way, a Faustian bargain. In the smelter's furnace, they encountered the source of the sulphur that had destroyed their crops, and were condemned to the most hellish labours. Some did not adapt to factory discipline and quit, while others toiled on, occasionally finding relief in the observance of Saint Monday (or its shift-worker equivalents). 'They're all dead' is the laconic answer I often received when asking about Puchuncaví residents who had worked in ENAMI Ventanas in the early years.

Mario Cisterna, a Puchuncaví worker who was imprisoned following the coup, recalled: 'My first job was agriculture. Then, when that business started there [ENAMI Ventanas], we sowed but didn't harvest . . . The factories, the smoke, wrecked the harvests and we had to look for something else. So I started working in construction, at 16, I was loading trucks to fill the ground for the factories . . . There, I learnt about the unions and got involved in political stuff. So they blacklisted us, we looked for jobs and couldn't find any, until when Salvador Allende's government came about . . . We worked in the smelter, with all the smoke, so that we'd die afterwards. The best at this were those who had worked in the fields, the most brute, because we were all used to physical effort. The gentlemen from the city were not for the hammer. Without us, those fellas [*weones*] would've been dead.'

The main UP parties – Partido Socialista de Chile (Socialist Party of Chile), Partido Comunista de Chile (Communist Party of Chile) and Movimiento de Acción Popular Unitaria (Popular Unitary Action Movement, left-Christian) – all had factory groups in Ventanas. José Carrasco, president of the Ventanas workers' association, was a Communist

86 Sanford Malman, Francisco Sabatini and Guillermo Geisse, 'El trasfondo socioeconómico del conflicto ambiental de Puchuncaví', *Ambiente y desarrollo* 9(4), 1995, 49–58; Josefina Buschmann and Daniela Jacob, 'Arqueología de una controversia: Reconstrucción histórica del centro industrial de Ventanas', thesis, Pontificia Universidad Católica de Chile, 2012, 28–31.

Party member. The extra-parliamentary Movimiento de Izquierda Revolucionaria (Revolutionary Left Movement, MIR) also had a member on the association's board, Guillermo Sotomayor, also known as Caballo Loco (Crazy Horse). Nonetheless, the centrist Partido Demócrata Cristiano (Christian Democratic Party) kept a large following, and the debate among the workers was thus lively and sometimes harsh. However, when on 11 July 1971 the Chilean Congress unanimously approved the nationalisation of all large-scale copper mines, support was broad among the population, including Ventanas's workers.[87] Allende defined copper as 'the wage of Chile', which pointed to a programme of state-led rent appropriation to finance individual and social wage increases, turning rents into wages, raising the value of labour power in the long run.

Internal noxiousness was acutely felt at ENAMI Ventanas, as regularly decried by *La unidad* since its very first issue of October 1969. The workers denounced a long list of health and safety hazards: toxic gases and powders containing sulphur and arsenic, extreme temperatures, heavy weights, dangers of falling, ergonomic injuries, long hours, etc. Before the coup, preventive measures and protective equipment were gradually improved, but severe gaps remained. In 1972, ENAMI set up a national bipartite health and safety committee, which drafted a scathing report and stressed the urgency of reform.

External noxiousness was discussed less frequently on the workers' journal, but a retired female white-collar employee had the following memories: 'You won't believe it, but people were very aware about the environment in those times, because they saw the impacts of sulphur pollution, and the workers had raised the issue. We wanted filters, new technologies, many things, even at that time! But then with the dictatorship we couldn't do it.'

Pollution, in fact, was to be reduced through the expansion and upgrading of the complex, as part of the governmental strategy to further industrialise the recently nationalised copper reserves. By early 1973,

87 Angela Vergara, *Copper Workers, International Business, and Domestic Politics in Cold War Chile*, University Park: Pennsylvania State University Press, 2008, 155–77.

ENAMI had prepared a plan featuring the installation of a sulphuric acid plant to capture part of the sulphur dioxide emissions, a flash smelting system and other more up-to-date technologies. *La unidad* noted that the acid plant would mitigate the 'massive amounts of sulphurous gases that are currently dispersed in the atmosphere, causing serious problems to the farmers and the population in general'.[88]

However, factory life was soon caught up in the troubles caused by the US-backed economic boycott and coup preparations, and intra-left divisions on how to respond. While Allende attempted in vain to strike a deal with the Christian Democrats, militant workers scrambled to prepare for the coup. Some Ventanas employees made political contacts with the *pirquineros*, the small miners whose copper ENAMI was tasked to buy and process, as they were mostly left-leaning and, more importantly, regularly handled dynamite as part of their job. Yet, eventually, the anti-coup workers were faced with the abysmal asymmetry of firepower between the labour movement and the armed forces, as witnessed by former ENAMI Ventanas employee and then political prisoner Rafael Maldonado: 'We held meetings in the factory itself, and people were desperate because they saw the coup coming but didn't know what to do . . . The electricity pylons in Valparaíso were being constantly sabotaged [by the opposition], with the city remaining in the dark. So we used to guard ENAMI's transmission towers and, among all of us, we didn't have one rifle! . . . The army knew very well what to do, they cut the telephones and attacked [ENAMI Ventanas] from the beach. And met no resistance. They occupied the factory in five minutes.'

The military takeover brutally quashed any open form of labour organising, and Ventanas was no exception. Interviewees estimated that hundreds of Ventanas workers were laid off in the aftermath of the coup, chiefly based on political criteria. Of these, many were imprisoned and tortured in different detention centres. For example, the Navy stationed a coal bulker ship called *Lebu* in Valparaíso Bay and used it as a secret detention and torture hub.[89] Tens of Ventanas workers were amassed

88 *La unidad*, April–May 1973, 12.

89 Gilberto Hernández (ed.), *El siniestro barco Lebu*, self-published, 2021.

there together with hundreds more political prisoners. Once released, some went into exile, while others faced a life of deprivation on the blacklists of Pinochet's regime.

The neoliberal dictatorship oversaw Chile's employment deindustrialisation, with the double result of erasing the strongholds of labour militancy and accommodating the appetite of global capital for minerals and monocrops.[90] On the surface, employment deindustrialisation in Chile indicates a convergence with Global North countries. For example, according to figures by the Groningen Growth and Development Centre, the share of manufacturing employment in the US declined from 24.4 per cent in 1954 to 8.7 per cent in 2010. Chile's trajectory is no different, having moved from 24.6 per cent to 9.6 per cent over the same period. However, if one considers export compositions, the persisting divergence is stark. Chile's economic complexity index – as calculated by the Observatory of Economic Complexity – is at −0.2, mainly due to its overwhelming reliance on raw and quasi-raw copper exports. The US's, instead, stands at 1.56, with machines and chemicals as its top export categories. Concurrently, environmental degradation under the dictatorship also gained ground with few obstacles.[91] While Chile fared well economically compared to the other South American countries, mainly because of rents stemming from high copper demand, the civic-military dictatorship entrenched high inequality and extractivism.

Pinochet's coup meant that ENAMI Ventanas's upgrading was postponed for almost twenty years, until the democratic transition of 1988–90. For decades, the plants operated without an emission capture system, and by the 1980s the environmental degradation of the area – caused by sulphur oxides, particulate matter and metals such as arsenic, copper,

90 Rayen Quiroga Martínez, 'La sustentabilidad socioambiental de la emergente economía chilena entre 1974 y 1999: Evidencias y desafíos', in Emir Sader (ed.), *El ajuste estructural en América Latina: Costos sociales y alternativas*, Buenos Aires: CLACSO, 2001, 255–74.

91 Gustavo Lagos and Patricio Velasco, 'Environmental Policies and Practices in Chilean Mining', in Alyson Warhurst (ed.), *Mining and the Environment: Case Studies from the Americas*, Ottawa: IDRC, 1999, 101–36.

lead, mercury and cadmium – was extreme.[92] Fenceline community protests against pollution and its denial therefore made a comeback during the last years of the dictatorship.[93] Moreover, in 1989, the press exposed the case of the so-called 'Green Men', cancer-struck Ventanas workers whose organs had changed colour because of exposure to noxious powders in the factory.

Ventanas's first sulphuric acid plant only began operating in 1990. After that, successive rounds of investment significantly reduced Ventanas's sulphur and particulate matter emissions, although its environmental performance stayed below the best international standards. Even though the 1990s indisputably marked a break with the unrestricted pollution that had characterised the former decades, part of the local population remained discontented. In fact, the post-dictatorship period witnessed a whopping expansion of noxious facilities in the area, including hydrocarbon, chemicals and asphalt terminals, the enlargement of the industrial and commercial port, a gas-fired power plant and two additional and highly polluting coal-fired power stations.[94] The new infrastructure brought significant hazards, noise, traffic and emissions (particularly nitrogen oxides and volatile organic compounds), while the leaks of coal, oil and metals into the sea further compromised artisanal fishing. Despite scientific publications pointing to severe pollution-related health risks, as well as widespread worries over these threats, no encompassing epidemiological study was conducted at the time.[95]

92 Eugenia M. Gayo et al., 'A Cross-Cutting Approach for Relating Anthropocene Environmental Injustice and Sacrifice Zones', *Earth's Future* 10, 2022, e2021EF002217.

93 Francisco Sabatini, Francisco Mena and Patricio Vergara, 'Otra vuelta a la espiral: El conflicto ambiental de Puchuncaví bajo democracia', *Ambiente y desarrollo* 12(4), 1996, 30–40.

94 Renzo Peragallo Díaz, 'La producción estatal de las zonas de sacrificio en Chile: Un estudio en profundidad del caso de Quintero-Puchuncaví', thesis, Universität Heidelberg & Pontificia Universidad Católica de Chile, 2020.

95 See, e.g., Eva Madrid et al., 'Arsenic Concentration in Topsoil of Central Chile Is Associated with Aberrant Methylation of P53 Gene in Human Blood Cells: A Cross-Sectional Study', *Environmental Science and Pollution Research* 29, 2022, 48250–9.

On 23 March 2011, a severe sulphur leak from the industrial complex, which had been bought in 2005 by Chile's main state-owned mining company, Corporación Nacional del Cobre de Chile (National Copper Corporation of Chile, CODELCO), left around thirty elementary school students and teachers from the nearby Escuela La Greda in need of medical attention. The mood for protest was now ripe. A coalition of local associations and schools promoted a campaign, and filed a legal complaint, demanding the closure of the Ventanas smelter. The matter was eventually settled with financial compensations, the relocation of the school and a clean production agreement. Nonetheless, smaller incidents of this kind continued, and on 24 September 2014 a tanker accident led to a massive 38,700-litre oil spill in the Quintero Bay.

On 21 August 2018, around seventy people, again mostly children, needed medical attention due to a massive peak in air pollution. Similar episodes continued in the following weeks, so that at least 1,400 people sought medical attention. The causes of the accidents were unclear, although in 2023 three managers of Chile's Empresa Nacional del Petróleo (National Petroleum Enterprise, ENAP) were found guilty over the transport of dangerous substances (sulphur-rich crude oil) in Quintero Bay. These accidents spurred a cycle of mobilisations against pollution, including an open-ended occupation of Quintero's square. A prominent role in this movement was taken by the group Mujeres de Zonas de Sacrificio en Resistencia (Women of Sacrifice Zones in Resistance, MUZOSARE) and by the artisanal fishers' unions, particularly Quintero's S24 Union. Established in 2016, MUZOSARE stressed the connections between extractivism, productivism and the patriarchy, and how women are on the frontline of the care work needed to address the effects of noxiousness.[96] The S24 Union had emerged in 2014 during the fishers' movement over the great oil spill, articulating an anti-capitalist discourse focused on territorial and food sovereignty. One of its

96 Paola Bolados García and Alejandra Sánchez Cuevas, 'Una ecología política feminista en construcción: El caso de las "Mujeres de zonas de sacrificio en resistencia"', Región de Valparaíso, Chile', *Psicoperspectivas* 16(2), 2017, 33–42.

leaders, Alejandro Castro, died in unclear circumstances on 4 October 2018, becoming a widely known symbol of the protest.

Local tensions are probably not unrelated to the transformations of employment in the area. The combined share of Puchuncaví and Quintero residents employed in manufacturing decreased from 16.6 per cent in 1982 to 3.8 per cent in 2019, making Quintero-Puchuncaví a clear case of noxious deindustrialisation.[97] Productivity gains and outsourcing contributed to an absolute decline in the number of manufacturing workers residing locally from 1,158 in 1982 to 441 in 2019 – a fall whose impact on the labour market was magnified by population growth. These tendencies were compounded when CODELCO took over the plants and introduced new technologies and managerial styles, as well as more centralised and formalised recruitment procedures. While in 1973 ENAMI Ventanas had almost 1,800 direct employees, in 2022 they had become about 800, in addition to a fluctuating number of around 1,000 outsourced workers (including outsourced cleaning, catering, security, transport and logistics workers). In stark contrast, productive capacity had increased from 150,000 to 420,000 yearly tonnes for the smelter, and from 84,000 to 400,000 yearly tonnes for the refinery.

With the shedding of industrial jobs, local employment grew especially in precarious service sector jobs. This was driven by an increase in tourism, which offers seasonal work with a high degree of informality, as well as the neoliberal rise of outsourcing, which turned formerly in-house and relatively secure jobs into flexible and riskier contractor employment. Both sectors are associated with a surge in the floating population. The lively Chilean *contratista* movement – which, in Ventanas, has seen important mobilisations by the copper workers since the mid-2000s and the dockers since the mid-2010s – did successfully mitigate the asymmetries between contractor and in-house employment, but significant disparities remain.[98] As discussed above, wage differentials act as a divisive wedge within the

97 Calculated by the author based on Chilean Census and Internal Fiscal Services data.

98 Fernando Durán-Palma and Diego López, 'Contract Labour Mobilisation in Chile's Copper Mining and Forestry Sectors', *Employee Relations* 31(3), 2009, 245–63.

working class, and were exacerbated by such trends. The overall result is an employment and residential fragmentation of the working class.

The effects of employment precarity were aggravated by Chile's weak welfare state. The targeted donations offered by the industries were deemed insufficient by the interviewees, and community activists also saw them as prone to take clientelist slants. Most participants noted how, despite the large production of wealth and pollution in the area, fence-line communities are deprived of basic services. In 2020, 32.1 per cent of the Puchuncaví population lacked access to basic services, compared to a national average of 13.8 per cent. Additionally, the multidimensional poverty rate in both Puchuncaví and Quintero was at about 27 per cent in 2017, compared to a national average of 20.7 per cent.[99] As often occurs, income and residential proximity to the polluting facilities are inversely proportional.[100]

The decoupling between the negative impacts of noxious production and its benefits turned Ventanas into an emblematic 'sacrifice zone', with the phrase gaining popularity in Chile's public debate (although the expression has also been criticised given its potential for stigmatisation). Nonetheless, such inequalities do not only apply on a local or national scale. They also reflect the global socioecological hierarchy analysed above. As local activists sharply pointed out, Ventanas's industrial area is better seen as an 'extractivist area', because even its factories are subordinated to Chile's role as a raw or quasi-raw copper exporter, with almost all of Chile's red metal further processed elsewhere.[101]

Today, Chile is by far the world's largest copper miner (26.3 per cent of global production) and the country with the largest known reserves

99 'Puchuncaví: Reporte comunal 2021', Biblioteca del Congreso Nacional de Chile, 2021, bcn.cl; 'Quintero: Reporte comunal 2021', Biblioteca del Congreso Nacional de Chile, 2021, bcn.cl.

100 Patricio Herrera, José Rojo and Valeria Scapini, 'Relationship Between Pollution Levels and Poverty: Regions of Antofagasta, Valparaíso and Biobío, Chile', *International Journal of Energy Production and Management* 7(2), 2022, 176–84.

101 MUZOSARE, *Feminismo popular y territorios en resistencia: La lucha de las mujeres en la zona de sacrificio Quintero-Puchuncaví*, Santiago de Chile: Heinrich Böll Stiftung Cono Sur, 2020, 61.

(200 million metric tonnes).[102] The 'commodity supercycle' deepened Chile's dependence on mining exports, particularly copper concentrates, which in less than twenty years saw a spectacular rise of 148 per cent in their share of total exports (Table 3.1).

Table 3.1. Chile's export of goods
(percentage of total prices of exported goods in USD FOB)

	2003	*2005*	*2007*	*2009*	*2011*	*2013*	*2015*	*2017*	*2019*	*2021*
Mining	40.4	52.2	61.7	57.2	59.9	56.6	52.0	53.9	51.4	61.9
Copper	36.8	47.3	57.0	53.3	54.5	51.8	48.2	49.3	47.3	56.3
Cathodes	21.5	25.4	29.9	31.0	31.9	24.5	22.8	21.4	19.4	22.1
Concentrates	12.7	18.2	21.4	17.7	17.8	21.6	20.9	24.0	25.6	31.5
Other Mining	3.5	4.8	4.6	3.8	5.3	4.7	3.6	4.4	4.0	5.6
Farming, forestry & fishing	9.9	6.1	4.8	6.6	6.1	7.3	8.4	8.3	9.9	7.0
Manufacturing industry	49.8	41.7	33.5	36.2	34.0	36.0	39.6	37.8	38.7	31.1

Source: Calculated by the author based on Central Bank of Chile data.

Chile dominates copper extraction, but industrial processing is now mostly located in China at all stages. In 2022, China carried out 47 per cent of global smelting, 42 per cent of refining and 54 per cent of further usage, while Chile only refined 8.9 per cent of the world's copper.[103] Chinese competition, with its combination of higher technologies and lower wages, increased the minimum scale of profitable plants and pushed most Chilean smelters into the red, including CODELCO Ventanas. The smelter thus became essentially reliant on subsidies funded by rents stemming from CODELCO's copper concentrate exports.[104] China displaced Global North countries as the main raw

102 S&P Global, *The Future of Copper: Will the Looming Supply Gap Short-Circuit the Energy Transition?*, cdn.ihsmarkit.com, 2022.

103 Ibid.

104 See Kevin Pérez et al., 'Environmental, Economic and Technological Factors Affecting Chilean Copper Smelters: A Critical Review', *Journal of Materials Research and Technology* 15(2), 2021, 213–25; Gabriel Rivas Castro and Juan Kornblihtt, 'Apropiación por el capital individual y en su conjunto de la renta de la

copper importer remarkably quickly, given that in 1995 it only refined 8.9 per cent of the world's copper. However, final manufactured products containing copper are still made by Asian low-wage workers to be disproportionately exported to high-wage countries.[105] This means that the latter are still the main beneficiaries of the current international division of labour and noxiousness.

The interviewed actors had diverging views on how to address noxiousness and employment deindustrialisation in an extractivist context. The copper workers' unions embraced a strategy that could be called 'sovereign sustainable industrialisation', in which state-led endogenous technological development would push Chile further up the commodity chain of the 'metal of electrification', reducing pressures to extract raw materials at a fast rate or to process them with outdated machinery.[106] For example, a Ventanas union representative said: 'I'm not for working at any cost. I also declare myself a defender of the environment, I don't want the damage to continue, I live here with my daughter and grandchildren, so it'd be an abomination to say I'm not interested. It's all about investing in environmental responsibility . . . If you want to progress with renewable energy, you'll need copper most! Copper will run out and this country is not smelting and refining it, it just delivers the concentrate. Chile can comply with the carbon reduction agreements only if it stops shipping the concentrate, smelting all the copper here.'

In contrast, community organisations, such as Mujeres por el Buen Vivir (Women for Buen Vivir), challenged the very idea of the expanded

---

tierra minera en Chile (1990–2017)', in Javiera Rojas Cifuentes et al. (eds), *La cuantificación del desarrollo histórico del capital en América del Sur: Estudios de largo plazo sobre la tasa de ganancia y la renta de la tierra, metodología y resultados*, Santiago de Chile: Ariadna, 2023, 92–136.

105 Yang Yu, Kuishuang Feng and Klaus Hubacek, 'China's Unequal Ecological Exchange', *Ecological Indicators* 47, 2014, 156–63.

106 See Gabriel Palma, 'Deindustrialisation, Premature Deindustrialisation and a New Concept of the Dutch Disease', *Revista NECAT* 3(5), 2014, 7–23; Gabriel Salazar, *En el nombre del poder popular constituyente (Chile, Siglo XXI)*, Santiago de Chile: Lom, 2011.

industrialisation of natural resources.[107] Activist Marta Aravena declared: 'If we change our vision of how we develop life, we'll start consuming less, there will be less companies, less everything . . . We need to build hope in the sense of teaching that happiness is in what people have. It's not brutally working all day long to amass a bunch of stuff, it's developing and enjoying your life.'

These perspectives are not totally irreconcilable. They might rather constitute, to use Álvaro García Linera's phrase, a 'creative tension' about what to prioritise.[108] In Ventanas, these tensions bubbled over the fate of CODELCO's smelter, with the copper workers' unions asking for further investments in cleaner technologies and several community organisations demanding its closure. After a new episode of mass contamination occurred on 6 June 2022, however, CODELCO and the government announced the smelter's gradual decommission, along with plans to relocate all direct employees to other CODELCO plants, assist outsourced workers in finding alternative jobs and build a larger 'mega-smelter' elsewhere in the country.

Copper was added to the US list of critical minerals in November 2025. This decision has nothing to do with the ecological transition, a policy objective that has been roundly rejected by Trump's white plan of capital. It must rather be understood in the framework of US attempts to defend its privileged position in the international division of labour, in contrast to the aspirations of both my union and environmentalist interviewees. Thea Riofrancos recently wrote that, historically, 'labelling resources as "critical" has justified government support for extraction and access, deregulation of safeguards, and a preference for strong-arm tactics over co-operation'.[109] In other words, the politicisation of nature

107 See Eduardo Gudynas, 'La dimensión ecológica del Buen Vivir: Entre el fantasma de la modernidad y el desafío biométrico', *Revista OBETS* 4, 2009, 49–53; Soledad Varea and Sofia Zaragocin (eds), *Feminismo y buen vivir: Utopías decoloniales*, Cuenca: Paydlos, 2017.

108 Álvaro García Linera, 'Las tensiones creativas de la revolución: La quinta fase del proceso de cambio', *Cuadernos FLACSO* 7, 2011, 1–31.

109 Thea Riofrancos, 'The "Critical Minerals" Rush Could Result in a Resource War', *Financial Times*, 12 March 2025.

that accompanies 'criticality' implies the exercise of extra-economic pressures that go beyond those already present in ordinary market relations. The threat of violence can therefore be seen as an implicit dimension of the critical mineral concept.

To contain the risks of further military escalation that this entails, pressure from below is needed – one capable of redirecting Global North policy away from the defence of white privilege and towards the containment of material throughput, via the provision of public and common goods and services that minimise waste, a reduction of inequalities and a transition to a more balanced international division of labour and noxiousness. This is not about ensuring a 'fair share' of wages and pollution to all. Rather, it is a step towards reducing the global intra-working-class stratification that stands in the way of the decommodification of production and nature, a decommodification needed to really address the ecological crisis.

4

Against Noxiousness: Working-Class Environmentalism from the Hidden Abodes of Production and Reproduction

Italian cinema director Tinto Brass once participated in a demonstration for the legalisation of sex work. On that occasion, feminist and prostitution abolitionist activist Elvira Banotti jumped on stage 'and punched me, while her comrades bombarded me with acorns screaming: "You pig!"', according to the man himself.[1] Before turning to erotic provocations and thus becoming a legitimate target for feminist acorns, Brass directed the film *Chi lavora è perduto* (*Who Works Is Lost*), released in 1963. The movie features a young Venetian man trying to escape work, sometimes finding a distraction in attending the Porto Marghera workers' strike pickets with his communist friends: 'They always told us work ennobles man [*sic*], they even wrote it on Auschwitz's gate, work makes us free. Of course, no doubt, right . . .'

Seven years later, director Dino Risi released *La moglie del prete* (The priest's wife), starring Sophia Loren and Marcello Mastroianni. The setting is still in Veneto, this time Padua, which surrounds the female protagonist with the thick hypocrisy of the patriarchal morals

1 Corriere del Mezzogiorno, 'Tinto Brass: "Le femministe? Mi tiravano le ghiande, la Banotti mi diede un cazzotto"', *Corriere del Mezzogiorno*, 31 January 2013, corrieredelmezzogiorno.corriere.it.

incumbent in a provincial and still deeply Catholic region. The young woman is determined to marry a priest with whom she has fallen in love, never sure whether her skirt is too short or too long relative to his ecclesiastical gown. The contradiction of challenging the clerical dogma of celibacy while still embracing the institution of marriage is, in a way, symbolic of the turbulent changes that Padua, an ancient university town shaken by 1968, was undergoing at the time.

These two films set the scene for both the theoretical and empirical gist of this chapter. In 1971, a few months after *La moglie del prete* hit the cinemas, Mariarosa Dalla Costa, based at the University of Padua, led a split from Potere Operaio to found Lotta Femminista (Feminist Struggle).[2] The latter would become the Italian branch of the Wages for Housework campaign, which also featured a different stance on sex work relative to Carla Lonzi and Elvira Banotti's Rivolta Femminile (Female Revolt) prohibitionist position, stressing instead the organisation and demands by sex workers themselves.[3]

Veneto *operaismo* and its renegade feminist line originated precisely between Padua and Venice. Their focal point was the industrial area of Porto Marghera, one of the most significant manufacturing hubs for Italy's post–Second World War labour movement, located in the mainland of the Venice council area.[4] In the early 1960s, this heart of '*laboratorio Veneto*' was at the centre of the encounter between intellectuals and students – led by then PSI's Padua city councillor Toni Negri – and militant workers disaffected with the line of the PCI and its associated union, Confederazione Generale Italiana del Lavoro (Italian General Confederation of Labour, CGIL).[5] Negri described his experience in Porto Marghera as 'an extraordinary decade of apprenticeship in class

2 Anna Curcio, 'Marxist Feminism of Rupture', *Viewpoint Magazine*, 14 January 2020, viewpointmag.com.

3 Louise Toupin, *Wages for Housework: A History of an International Feminist Movement, 1972–77*, London: Pluto Press, 2018.

4 Cesco Chinello, *Sindacato, Pci, movimenti negli anni Sessanta: Porto Marghera-Venezia, 1955–1970*, Milan: FrancoAngeli, 1996.

5 Devi Sacchetto and Gianni Sbrogiò (eds), *Quando il potere è operaio: Autonomia e soggettività politica a Porto Marghera, 1960–1980*, Rome: Manifesto Libri, 2009.

struggle', while Dalla Costa 'would go to leaflet in Porto Marghera, in a pale dawn full of mosquitoes, discovering what a factory is, its rhythms, its health hazards, and its history'.[6]

While this setting functioned as an incubator for influential intellectuals, the theories produced by Porto Marghera's workers themselves have been largely forgotten. The perspective developed by these factory militants and their associates, which together formed the Porto Marghera *operaista* group, is, to my knowledge, the first systematic application of *operaismo* to ecological politics.[7] They addressed such issues from the late 1960s onwards, when the *operaistas* had a determining influence in Porto Marghera's class struggles, spurred by the fiendish toxicity and hazards they faced in the local factories, particularly Montedison's Petrolchimico.[8] Under the impulse of Petrolchimico technician Augusto Finzi, the Porto Marghera *operaista* group's original contribution was grounded on a thesis of the inherent noxiousness of capitalist work. This eventually led to the proposal of a counterpower able to determine 'what, how and how much to produce' based on common needs that encompassed the environment, pointing to the prospect of struggling for a different, anti-capitalist technology compatible with the sustainable reproduction of life on the planet.

In the 1960s and '70s, Porto Marghera's struggles against noxiousness originated from workplace grievances, as part of a broader national and international pattern. In the 1990s and 2000s, by contrast, the fulcrum of action shifted to the surrounding communities, with workplace-centred organisations acting in response. In such a context, the community-centred and workplace-centred working-class environmentalist camps

6 Antonio Negri, 'Un intellettuale tra gli operai', in Sacchetto and Sbrogiò, *Quando il potere è operaio*, 140–50, 140; Mariarosa Dalla Costa, *Women and the Subversion of the Community: A Mariarosa Dalla Costa Reader*, Oakland, CA: PM Press, 2019, 161.

7 This group changed denomination many times, using simultaneously different names for different purposes (Potere Operaio, Comitato Operaio, Comitato Politico, Assemblea Autonoma di Porto Marghera, Lavoro Zero, Controlavoro, Collettivo di Lotta contro le Produzioni Nocive, etc.). Here, it is referred to simply as the 'Porto Marghera *operaista* group' for the sake of brevity.

8 Gilda Zazzara, *Il Petrolchimico*, Padua: Il Poligrafo, 2009.

diverged over Petrolchimico's chlorine-based production, with the former demanding a just transition away from chlorine and the latter a just transition within it. To understand such a shift, this chapter builds on the renegades of *operaismo* feminism to lay the bases for an 'ecological turn' in class composition analysis, which allows us to see a class dimension in both workplace- and community-centred environmentalist struggles. Workplace-centred struggles are conflicts over the conditions under which workers produce commodities or reproduce labour power, while community-centred struggles are conflicts over the conditions of workers' own reproduction. As further argued in the conclusion, the convergence between workplace and community mobilisations is a critical step in the construction of alternatives to the jobs versus environment dilemma.

An ecological turn in class composition analysis

As seen in the introduction, class composition analysis is a method of inquiry aimed at identifying the most relevant lines of intra-working-class division to inform political interventions geared towards overcoming them. And, as discussed above, the wage hierarchy is a vector of working-class stratification and decomposition on a global level, which contributes, in turn, to an ecological hierarchy that results in differential levels of exposure to capitalist noxiousness. To deepen our understanding of the ecological dimension of working-class composition, I propose a threefold conceptual expansion: 1) a conception of the working-class based on dispossession, rather than exploitation; 2) a conception of work that includes both production and reproduction; and 3) a conception of working-class interests that encompasses both workplace and community.

First, class. As argued in Chapter 2, if the capitalist class relation is understood as grounded in the fundamental separation between workers and means of production, then class should be understood as arising from dispossession rather than exploitation. The working class is not just made of those who are directly engaged in commodity production for a wage. The working class, in all its heterogeneity, is made of all those who lack ownership and control of significant magnitudes of means of production, including reproductive and surplus workers. This is not an expansion relative to the late Marx, who saw the surplus population as a segment of the working class, but it took the perspectives of Marxist feminists and Marxists from the Global South to fully draw out its implications. The conceptualisation embraced here, even with its exclusion of the middle class, is nonetheless an expansion relative to the narrow notions of the working class dominant across much cultural representation, as well as within the labour movement itself.

With the relative numerical and political downsizing of what, for some, is the ideal-typical worker – white, male, blue-collar, much like *operaismo*'s mass worker – the struggles waged by different working-class segments have again practically raised the theoretical need for a broadened understanding of the class. As Verónica Gago put it:

> Feminisms, through the strike, contest the boundaries of what is defined as labor and, therefore, the working class, opening the category back up to new experiences and demonstrating its historically exclusionary meaning. Even more: the strike broadens feminist experiences, taking them to spaces, generations, and bodies not included in earlier feminist practice. Moreover, it points to something beyond the 'patriarchy of the wage' and its heteronormative rule.[9]

Second, as first argued by pioneering Marxist feminist Mary Inman, capitalist work encompasses all activities – waged and unwaged, directly productive and reproductive – subordinated in obvious or hidden ways to capital accumulation, no matter the economic sector.[10] The dispossessed, in fact, work either in the making of commodities, as in directly

9 Verónica Gago, *Feminist International: How to Change Everything*, London: Verso, 2020 [2019], 14.

10 Mary Inman, *In Woman's Defense*, Los Angeles: Committee to Organize the Advancement of Women, 1940; see also Maria Mies, *Capital and Accumulation on a World Scale: Women in the International Division of Labour*, London: Zed Books, 2014 [1986].

productive work, or in the non-directly-commodified making of an employable workforce for capital, as in reproductive work.

Different approaches to reproductive work have been elaborated recently.[11] The relationship between reproductive workers and capital is understood here to be indirect, because it is mediated by a waged household member (as in housework), or by the state (as in the provision of non-commodified welfare services). The mediators acquire money from capital through wages or taxes and distribute it to reproductive workers in exchange for their labour. The split between production and reproduction thus establishes an intra-working-class hierarchy between waged and unwaged (or, rather, indirectly waged) household members.

The distinction between directly productive and reproductive work is determined by the 'line of value', or the frontier of commodification.[12] That is, by the social form under which work takes place, not by the type of concrete activity. For example, food is necessary to the reproduction of the workforce. Yet, while growing cash crops for an agricultural company is directly productive of value, cultivating food for self-consumption within a capitalist context is, by contrast, reproductive.

On this basis, a further distinction can be made between reproductive work, defined by social form, and care work, as a type of concrete labour. Reproductive work is the non-directly commodified making of labour power for capital. Care work refers to concrete tasks that involve attending to the physical and emotional needs of people, for example, elder care. As a material content, care work can take different social forms. It can be directly productive, if it is performed as a service that companies sell on the market, or reproductive, if it is delivered by household members or as a decommodified welfare service. The neoliberal phase of

11 See, e.g., Tithi Bhattacharya (ed.), *Social Reproduction Theory: Remapping Class, Recentering Oppression*, London: Pluto Press, 2017; Susan Ferguson, *Women and Work: Feminism, Labour, and Social Reproduction*, London: Pluto Press, 2019; Alessandra Mezzadri, 'A Value Theory of Inclusion: Informal Labour, the Homeworker, and the Social Reproduction of Value', *Antipode* 53(4), 2021, 1186–205; Cristina Morini, *Vite lavorate: Corpi, valore, resistenze al disamore*, Rome: Manifesto Libri, 2022.

12 Leopoldina Fortunati, *The Arcana of Reproduction: Housewives, Prostitutes, Workers and Capital*, London: Verso, 2025 [1981], 119.

capitalism has seen an increase in the participation rate of the female workforce in waged employment and a related commodification of care work.[13] As a result, in a significant share of households, especially in the Global North, care work is carried out by low-waged, often racialised women external to the family.[14]

Third, working-class interests are seen here as related to both workplace and community. Similar to the distinction between productive and reproductive work, the distinction between workplace and community is not based on concrete, physical boundaries, but on social relations. The workplace is the domain of 'workers-as-producers-or-reproducers', while the community is the domain of 'workers-as-reproduced'. In some cases, a physical space is both a workplace and a community setting for the same people. For example, the home is both a workplace for reproductive work (or for productive work too, as in commodified remote working) and a community setting. In other cases, a physical space is a workplace to some and a community setting to others. For example, a hospital is the workplace of its employees and a community setting for its patients.

Since at least the publication of the English version of Mariarosa Dalla Costa's seminal essay 'Women and the Subversion of the Community', the sphere of workers-as-reproduced has been translated as the 'community' in English debates.[15] It is worth noting, however, that in the Italian political jargon the usual term is 'territory'. In fact, the renegade *operaismo* feminists had critically re-elaborated Mario Tronti's thesis of the 'social factory', according to which capitalist production tends to subordinate all spheres of social life to the profit imperative, particularly the 'territory'.[16] As Silvia Federici wrote:

13 Clare Ungerson, 'Social Politics and the Commodification of Care', *Social Politics* 4(3), 1997, 362–81.

14 Sara R. Farris and Sabrina Marchetti, 'From the Commodification to the Corporatization of Care: European Perspectives and Debates', *Social Politics* 24(2), 2017, 109–31.

15 Mariarosa Dalla Costa and Selma James, *The Power of Women and the Subversion of the Community*, Bristol: Falling Wall Press, 1972.

16 Mario Tronti, *Workers and Capital*, London: Verso, 2019 [1966].

> Tronti referred here to the increasing reorganization of the 'territory' as a social space structured in view of the needs of factory production and capitalist accumulation. But to us, it was immediately clear that the circuit of capitalist production, and the 'social factory' it produced, began and was centered above all in the kitchen, the bedroom, the home – insofar as these were the centers for the production of labor-power.[17]

The term 'territory' has an advantage over 'community' as it points more directly to the connection between the conditions of workers' reproduction and their ecologies. Indeed, reproduction is not just social, it is socioecological, in the sense that the reproduction of labour power also depends on certain relations with non-human nature. This can be connected to Latin American feminist theorisations of the link between feminist struggles and the defence of the territory.[18] Similar to the concept of noxiousness, the notion of *cuerpo-territorio* (body-territory) helps avoid an ontological dualism between human health and environmental degradation by highlighting the dialectical co-constitution of the two, as 'what is done to the body is done to the territory and vice versa'.[19]

Working-class interests are often thought of as workplace-centred (secure jobs, high wages, short hours, safe conditions, etc.), and wealth redistribution via higher wages for shorter hours would indeed help to overcome the jobs versus environment dilemma by reducing the need for jobs in the first place. However, workers do not vanish after leaving their workplaces. Rather, they return to their neighbourhoods, breathe the air outside factories and offices and enjoy their free time by relating to their ecologies.[20] These communities are also stratified along class lines,

17 Silvia Federici, *Revolution at Point Zero: Housework, Reproduction, and Feminist Struggle*, Oakland, CA: PM Press, 2012, 7–8.

18 Astrid Ulloa, 'Feminismos territoriales en América Latina: Defensas de la vida frente a los extractivismos', *Nómadas* 45, 2016, 123–39.

19 Sofia Zaragocin and Martina A. Caretta, '*Cuerpo-Territorio*: A Decolonial Feminist Geographical Method for the Study of Embodiment', *Annals of the American Association of Geographers* 111(5), 2021, 1503–18, 1508.

20 Bue R. Hansen, 'The Interest of Breathing: Towards a Theory of Ecological Interest Formation', *Crisis and Critique* 7(3), 2020, 108–37.

intersecting with other hierarchical vectors. The community, then, is not seen here – in liberal terms – as the realm of formally equal citizens and consumers. As the community is also crossed by class relations, workers' interests do not lie solely in their conditions of work but also in their conditions of reproduction (e.g. price levels, welfare services, healthy ecologies). Workplace-centred struggles are then those waged for workplace-centred interests, while community-centred struggles are those fought for community-centred, reproductive interests (see Figure 4.1).

Figure 4.1 Workplace and community

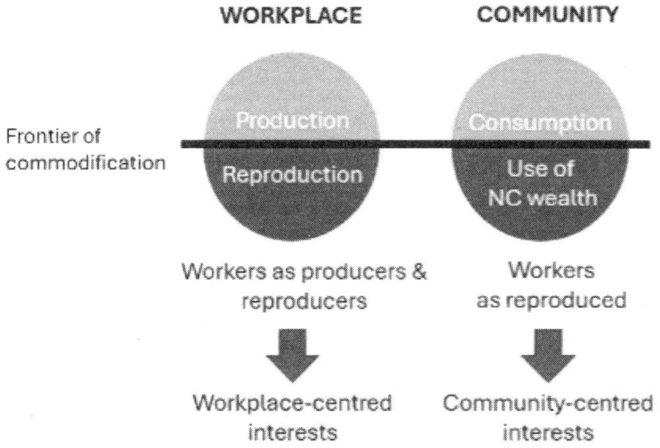

Source: Author. 'NC' stands for 'non-commodified'.

Under the frame just outlined, it becomes possible to assess Hardt and Negri's claim that, in neoliberal capitalism, the distinction between production and reproduction no longer holds:

> In the biopolitical context of Empire ... the production of capital converges ever more with the production and reproduction of social life itself; it thus becomes ever more difficult to maintain distinctions among productive, reproductive, and unproductive labor ... The progressive indistinction between production and reproduction in the biopolitical context also highlights once again the immeasurability of

> time and value. As labor moves outside the factory walls, it is increasingly difficult to maintain the fiction of any measure of the working day and thus separate the time of production from the time of reproduction, or work time from leisure time. There are no time clocks to punch on the terrain of biopolitical production; the proletariat produces in all its generality everywhere all day long.[21]

If one keeps the distinction between social form and concrete activity in mind, it is irrelevant whether work takes place within factory walls, under a ticking clock, or not.[22] Work productive of value makes commodities, and it is thus directly subjected to the whip of market competition. If such work takes longer than the average labour time, then it generates less than normal profits. It is up to the employer (or to the worker herself, in cases of self-employment) to ensure that the work is done quickly enough. This can be done in manifold ways, and the obligation to turn up in a physical workplace for a fixed number of hours is not always the most effective one, skiving off being a time-honoured working-class tradition. Piecework, deadlines, bogus self-employment and performance-conditioned fixed-term contracts are just some of the ways in which the law of value can be forced upon workers wherever they work, without the need to count directly how many hours and minutes they have been on the task.

Conversely, since reproductive work is not directly commodified, the amount of time spent on it is, on the one hand, the worker's own problem. On the other hand, however, reproductive work remains (indirectly) commodified, such that it is subordinate to commodity production and thereby internal to capital accumulation. If it does not reproduce an employable labour power for capital, this is eventually sanctioned with varying amalgams of unemployment, underemployment and super-exploitation.

21 Michael Hardt and Antonio Negri, *Empire*, Cambridge, MA: Harvard University Press, 2000, 402–3.

22 See Maya A. Gonzalez and Jeanne Neton, 'The Logic of Gender: On the Separation of Spheres and the Process of Abjection', *Endnotes* 3, 2013, 56–91.

Based on all of the above, I argue that working-class environmentalism can be both workplace-centred and community-centred. Workplace-centred environmentalist struggles demand a cleaner labour process from the workplaces, while community-centred environmentalist struggles reclaim a healthier and safer ecology in the neighbourhoods. But how is it to be determined whether a movement or an organisation are 'working-class'? The criterion cannot be the sociological composition of their leaders or activists, which is almost always a mix of working-class and middle-class people, as per the distinction laid out in Chapter 2. For example, labour confederations usually organise both middle-class and working-class employees, and even in industry there are often middle-class union activists who hold posts of responsibility within their firms. A movement is working-class, then, if its self-understanding is largely based on a working-class perspective *and* if this perspective is validated by a significant working-class constituency. Fenceline communities located in the vicinity of highly noxious facilities are usually characterised by lower-than-average incomes, as capitalist and middle-class households can more easily choose, or relocate to, cleaner areas.[23] Such low incomes are a proxy for a high working-class presence and thus an indicator of potential for community-centred working-class environmentalism.

As outlined in the introduction, working-class composition analysis encompasses three domains: technical composition, social composition and political composition. The technical composition refers to capital's organisation of the workers as labour power in the workplace.[24] The social composition is a recent innovation, designating the ways in which workers are reproduced as labour power in the community.[25] Finally,

23 Grettel Navas, Giacomo D'Alisa and Joan Martínez-Alier, 'The Role of Working-Class Communities and the Slow Violence of Toxic Pollution in Environmental Health Conflicts: A Global Perspective', *Global Environmental Change* 73, 2022, 102474.

24 Romano Alquati, 'Composizione della classe: Una ricerca sulla struttura interna della classe operaia italiana', *Classe operaia* 13(1), 1965, 8–14.

25 Seth Wheeler and Jessica Thorne, 'The Workers' Inquiry and Social Composition', *Notes from Below*, 29 January 2018, notesfrombelow.org.

the political composition constitutes the 'subjective' side of the workers as working class. The relationship between the objective (technical and social) and subjective (political) sides of class composition is not a determinist one, because there is a leap of agency and contingency between the two. As concisely phrased by Seth Wheeler and Jessica Thorne: 'Technical [and social] composition sets the basis for political composition, although the movement from one to the other is not mechanical or predictable.'[26]

For example, the rise of an electronics export industry in China has generated a distinctive working-class composition. In terms of the technical composition, a mostly 'low-skilled' workforce is deployed – following gendered and ethnic patterns – on assembly lines geared towards the flexible production of constantly changing models.[27] For the social composition, the most characteristic arrangement is the 'dormitory regime'.[28] Rural migrant workers on fixed-term employment are housed in collective dorms provided by their employing companies, which they have to leave once their job contract has expired. During their employment, their needs are mostly managed by their company. Only the hard core of tasks that are very difficult to commodify, such as personal hygiene or emotional care, are left to their own reproductive work. After they are laid off, the *hukou* legal system restricting internal mobility makes sure they go back to their original home. Their reproduction is now outsourced to the rural village, ensuring the permanent supply of an itinerant workforce, moving between towns and countryside according to the vagaries of global demand. In terms of the political composition, the absence of independent unions has resulted in the diffusion of wildcat strikes that sometimes turn into riots.[29]

26 Ibid.

27 Jenny Chan, Mark Selden and Pun Ngai, *Dying for an iPhone: Apple, Foxconn, and The Lives of China's Workers*, London: Pluto Press, 2020.

28 Pun Ngai, 'Gendering the Dormitory Labor System: Production, Reproduction, and Migrant Labor in South China', *Feminist Economics* 13(3–4), 2007, 239–58.

29 Eli Friedman, *Insurgency Trap: Labor Politics in Postsocialist China*, Ithaca, NY: ILR Press, 2014; Ching K. Lee, *Against the Law: Labor Protests in China's Rustbelt and Sunbelt*, Berkeley: University of California Press, 2007.

Introducing social composition into the equation makes explicit the analysis of the realm of workers-as-reproduced first theorised by the renegades of *operaismo* feminism. This allows us to see how the 'objective' side of working-class composition is bifurcated into the technical (workplace) composition and the social (community) composition. Two interrelated issues follow. First, the dialectical relationship between capital composition and class composition also concerns the social composition of the working class. Second, the social composition, just like the technical composition of the working class, is also fragmented and stratified.

Regarding the first point, Silvia Federici wrote:

> Only at the end of the nineteenth century did the capitalist class begin to invest in the reproduction of labour, in conjunction with a shift in the form of accumulation, from light to heavy industry, requiring a more intensive labour discipline and a less emaciated type of worker . . . In Marxian terms, we can say that the development of reproductive work and the consequent emergence of the full-time housewife were the products of the transition from 'absolute' to 'relative surplus' value extraction as a mode of exploitation of labour.[30]

We have seen in Chapter 1 how the general law of capitalist accumulation, according to which capitalist technological development tends to increase working-class precarity, polarises the technical composition of the working class into highly qualified workers (possessing the abstract skills necessary to supervise sophisticated machinic systems) and surplus workers (scratching an insecure income from a wide array of less complex tasks). This comes with a parallel differentiation in the social composition, seen in the varying quality of reproductive arrangements achieved by different global working-class segments.

As pointed out by Angela Davis among others, working-class reproduction has always been more diverse than the picture initially drawn by

30 Federici, *Revolution at Point Zero*, 94.

its theorists.[31] In fact, the early analyses of reproduction were centred on the ideal-typical white working-class family: the man a long-term factory worker and the woman a full-time housewife. Nonetheless, as Federici herself also noted, capital has happily and consistently exploited emaciated workers in those sectors where their health and good spirits do not constitute a boost to productivity.[32] This is what makes the difference between spacious houses connected to all utilities and slums deprived of even drinkable water, between the enforcement of decent health and environmental standards in 'respectable' neighbourhoods and their systematic violation or complete absence in the ghettos of the surplus working class.

The fracture between production and reproduction, as well as intra-working-class stratifications, also take on a spatial dimension. As argued above, the distinction between workplace and community is defined by social relations, not physical space. Nonetheless, just as the value-form shapes the ways in which concrete labour is carried out, it also has an impact on the spatial distribution of people and their activities. Structurally, some tasks and spaces tend to fall under direct production, while others fall under reproduction. Manufacturing tends to be done by directly productive workers in factories, while many activities related to raising children are more often kept outside of direct commodification and carried out in homes and public institutions. Similarly, there is a structural tendency for different working-class segments to live in different places and under different reproductive regimes. Phillip Neel's analysis of Tanzanian political economy provides a vivid example of the divergences of working-class social and technical composition, as well as their relation to the rising organic composition of capital:

> That same disjunction visible in the contrast between half-empty condo complexes rising over the city and the *uswahilini* [informal settlements] stretching out below is here replicated at the level of the economy at large: a few big firms tower over the others in terms of

31 Angela Y. Davis, *Women, Race and Class*, New York: Random House, 1981.
32 Federici, *Revolution at Point Zero*, 105.

investment and output, but the real bulk of employment lies in the sprawling, informal mass of handicrafts, services, and grey market subsistence.[33]

Fenceline communities living by highly polluting industries are often disproportionately composed of the most disadvantaged ranks of the working class (also racialised in many cases) and do not necessarily have widespread access to jobs in the factories. For these working-class segments, ecological transitions would mean a welcome drop in higher-than-average cancer rates and other diseases. Furthermore, as seen in Chapter 1, recent decades have seen the evolution of the organic composition of capital resulting in a relative 'deterritorialisation' of the workforce in capital-intensive industry and mining. Following the end of the era of company towns and expensive transport, the workers employed in polluting sites are now less likely to live in their immediate vicinity. Thus, to the workforce of such toxic workplaces, ecological transitions more likely represent the risk of ending up in more precarious and lower-paying jobs. This dynamic of deterritorialisation fuels the potential for intra-working-class tensions between those employed in polluting industries but living far from them, and those living nearby but working in other sectors.

Accordingly, *the rift between the technical and social compositions of different working-class segments means that workplace-centred and community-centred interests are not immediately aligned. Rather, they need to be recomposed at the political level.* The 'ecological turn' in class composition analysis proposed here sees the social composition as a 'door to the territory', to paraphrase Mariarosa Dalla Costa's notion of reproduction as a house with a 'door to the garden' of ecology.[34] Just as ground rent must be added to the wage–profit dualism, the social composition must be added to the technical–political composition

33 Phillip Neel, 'Broken Circle: Premature Deindustrialization, Chinese Capital Exports, and the Stumbling Development of New Territorial Industrial Complexes', *International Labor and Working-Class History* 102, 2023, 94–123, 106.

34 Dalla Costa, *Women and the Subversion of the Community*, 241.

dualism, turning the labour–capital dualism into a trinity encompassing nature too. In this way, class composition analysis can be used to assess how workers are fragmented in relation to environmental degradation, gauging their differential dependencies on and exposures to noxiousness, and informing platforms of demands aiming to articulate workplace and community interests together.

The threefold conceptual expansion of working class, work and working-class interests outlined above is meant to move beyond narratives that reinforce the jobs versus environment dilemma. If 'real' work is waged and industrial only, and thus the 'real' working class is disproportionately male (and white, until recently), and if 'real' working-class interests mainly consist in keeping one's job, then it is hard to see a way out. Even more so if community mobilisations are considered as devoid of any class content, as if the inhabitants of the mostly working-class communities affected by severe noxiousness did not have to work for a living. Conversely, an inclusive understanding of such concepts lends itself more easily to the building of coalitions among workers differentially located within the gender-race-class hierarchy.

Operaismo versus capitalist noxiousness

A group of workers march to Porto Marghera's Petrolchimico factory carrying a four-metre-tall wooden cross. The date is 27 February 1973; the place Venice. Yet the setting for this story is not what people normally imagine when thinking of the 'City of Water'. There are no gondolas, ancient bridges or winged lions in sight, only the tangle of pipes and chimneys typical of so many petrochemical sites across the world. When the marchers are almost at the factory gates, they lay the cross down and tie a mannequin to it, before raising it and pitching it in the ground. This is no Catholic ceremony, however. It is a protest against noxiousness. The puppet wears a gas mask and is a symbol of what these workers refuse to become: a human sacrifice on the altar of 'progress'.

Porto Marghera's industrial area was established from 1917 onwards, and significantly enlarged in the 1950s when the company Edison built its petrochemical plants there.[35] Edison merged with Montecatini in 1966 to become Montedison, the largest employer in the area with around 14,000 direct employees and 5,000 outsourced workers in the early 1970s.[36] The industrial cluster also included a port, power plants, metal factories, a shipyard and an oil refinery. It reached its peak of employment in the mid-1960s with about 40,000 workers, some of whom lived in the vicinity while many others commuted from several areas of Veneto.[37] Meanwhile, the population of Marghera, the residential area by the port, grew into a large working-class community.[38]

Under the supervision of former fascist officials, the 'production inferno' of the post-war decades featured appalling hazards. Exposure to heat, smoke and powders, and flimsy health and safety measures, resulted in widespread accidents and occupational diseases such as cancers, silicosis, asbestosis and hepatic and dermatological pathologies.[39] Montedison's Petrolchimico began manufacturing PVC using Monsanto patents that involved high toxicity. Montedison worker and then environmental activist Gabriele Bortolozzo wrote: 'The chloralkali units were sadly disastrous for the workers, due not only to chlorine but also to mercury vapour . . . To calm things down, all the company did . . . was distributing abundant doses of cow milk.'[40]

Up to the 1950s, Porto Marghera's working-class composition was centred on the figure of the 'peasant-factory worker' we encountered in Chapter 2, employed in factories but reproduced in the countryside. This was a time of labour weakness and relative social peace, as Veneto was a conservative region, where the left had limited influence. However, the

35 See Cesco Chinello, *Storia di uno sviluppo capitalistico: Porto Marghera e Venezia 1951–1973*, Rome: Editori Riuniti, 1975.

36 Gabriele Bortolozzo, *L'erba ha voglia di vita: Autobiografia e storia politica tra laguna e Petrolchimico*, Venice: Associazione Gabriele Bortolozzo, 1998, 61.

37 Gilda Zazzara, 'I cento anni di Porto Marghera', *Italia contemporanea* 284, 2017, 209–36, 224.

38 Alessandro F. Nappi, *Storia di Marghera: Da periferia a città*, Venice: CSC, 1994.

39 Gianni Moriani, *La nocività in fabbrica e nel territorio*, Verona: Bertani, 1974.

40 Bortolozzo, *L'erba ha voglia di vita*, 33.

situation began to change rapidly in the 1960s. A new cohort of young workers, many with a rural background too, showed their determination not to accept the hard fate endured by their parents, with the first major strike at Petrolchimico in July 1963.

The Porto Marghera *operaista* group originated in this context. Throughout the 1960s, it developed its own perspective centred on egalitarian demands as an instrument for working-class recomposition: equal wage increases (in absolute value) for everyone or inversely proportional wage hikes, a guaranteed minimum wage for all workers, the reduction of the working week, and an abolition of differences in statutory conditions (holidays, social contributions, etc.) among the different categories of employees. By 1967, the group had become an extra-parliamentary local organisation known as Potere Operaio, with an office in Marghera and a journal titled *Il potere operaio: Giornale politico degli operai di Porto Marghera*, leafleting regularly outside the factory gates.

The group's platform proved to be largely successful when the discontent of Petrolchimico's workers exploded in the summer of 1968.[41] By then, Potere Operaio's influence had extended to five out of the seven members of the Petrolchimico workers' representation body (*commissione interna*). The 1968 mobilisation was thus characterised by a 'dual leadership' in which Potere Operaio conducted a hard rank-and-file campaign of wildcat strikes, active pickets, mass marches and roadblocks, culminating in the occupation of the Venice-Mestre train station on 1 August 1968, while the unions were pressured to endorse the egalitarian demands of the base and negotiate for them at the institutional level.[42] Class conflict would also ride high throughout 1969, when it expanded countrywide with the 'Hot Autumn' strike wave over the renewal of the national-sectorial collective agreements.

The Porto Marghera group's theorising over noxiousness was spearheaded by Petrolchimico technician Augusto Finzi, with the assistance

41 See Marie Thirion, *Organiser le pouvoir ouvrier: Le laboratoire opéraïste de la Vénétie (1960–1973)*, Marseille: Agone, 2024; Pietro Trevisan, *Petrolchimico: Autobiografia di un sopravvissuto*, Verona: Cierre, 2017.

42 Mestre is the other urban conglomeration on the mainland side of the Venice council area.

of external intellectuals such as high school chemistry professor Lino Bassani, academic physicians Libero Battiston and Loriano Bonora, University of Padua's occupational health professor Bruno Saia, and Claudio Sossai, a mathematician for Italy's National Research Council. Born in 1941 to a well-off Jewish family based in insular Venice, Finzi spent part of his early childhood in a Swiss refugee camp to escape the Shoah, in which the German chemical industry played a key and dreadful role. In 1960, he received a high school diploma in chemistry from Mestre's Istituto Pacinotti – 'where the beginning and end of the classes were signalled by the sound of a [factory-like] siren' – and immediately began working at Petrolchimico.[43]

In many ways, Finzi's 'factory intellectual' profile was the counterpart to Italo Sbrogiò's practical and charismatic approach.[44] The Porto Marghera group's most prominent leader, Italo Sbrogiò was a Petrolchimico maintenance worker hailing from a poor and numerous rural family. Having started work before completing elementary school, he reached Porto Marghera aged sixteen as a precarious construction labourer. His early factory years were far from rosy: 'Chemical technology crashed on us like an avalanche, so much so that many labourers and technicians paid for this experience with their lives.'[45] Perhaps this contributed to his decision to join Potere Operaio, both through and despite his activism in the PCI and CGIL (from which he was expelled in 1969).

Up to the 1960s, the routine policy regarding noxiousness was its 'monetisation'. Unions and management would agree the danger money for the damage and risks that workers suffered. The first notable struggle over noxiousness involving the Porto Marghera group took place in

43 Interview with Augusto Finzi by Manuela Pellarin, 2002. See Manuela Pellarin's documentary *Gli anni sospesi: Movimenti e percorsi politici a Porto Marghera* (2009).

44 Not to be confused with his younger brother Gianni Sbrogiò, an employee of the metal factory AMMI and also a prominent member of the Porto Marghera group.

45 Italo Sbrogiò, *La fiaba di una città industriale: 1953–1993, 40 anni di lotte*, Venice: El Squero, 2016, 13–14.

1967, precisely when Montedison decided to withdraw the noxiousness bonus from some factory units. This was merely a reactive struggle to restore what could be called the 'moral economy of noxiousness', which had held sway up to that point. Nevertheless, it marked the beginnings of the group's interventions around noxiousness.

At that time, other Italian leftist organisations were beginning to analyse such issues.[46] For example, activists around medicine professor Giulio Alfredo Maccacaro founded the association Medicina Democratica (Democratic Medicine) with strong ties to the labour movement, as shown in the journal *Sapere* after Maccacaro became its director. Important figures in this milieu were Luigi Mara, a technician at Montedison's Castellanza (Varese) plant, and Ivar Oddone, a physician involved in struggles against noxiousness in FIAT's Turin factories and author of the influential 1969 union pamphlet 'The Work Environment'.

The specificity of the Porto Marghera group's conception was its linkage to the *operaista* refusal of work. In this perspective, as we know, capitalist work is the production of capital and thus the reproduction of a society of exploitation. The strategic aim of class struggle is therefore not understood as an affirmation of work as a positive value, but as negation. As Finzi recalled:

> The great, long-lasting, existential fracture with the union [CGIL] and the PCI was that they considered work as an instrument that shapes people, while the perspective of those who – like me – became workers at that time and faced the contradictions of the factory was that the human being comes before, and it is the human being who must decide the conditions of acceptability and liveability of work.[47]

The combination of the 'strategy of refusal' with the dire health and safety conditions faced by workers at the time led the Porto Marghera

46 Stefania Barca, 'On Working-Class Environmentalism: A Historical and Transnational Overview', *Interface* 4(2), 2012, 61–80; Giovanni Berlinguer, *La salute nelle fabbriche*, Bari: De Donato, 1969.

47 Interview with Augusto Finzi by Manuela Pellarin, 2002.

group to the core idea of capitalist work as inherently noxious.[48] The reduction of worktime was thus the commended tactic to transform the fight for survival in the factory into a struggle for liberation from capitalist work: 'We know, comrades, that the only real way to reduce noxiousness to a minimum is *spending less time in the factory*. We thus demand: more holidays, less worktime, more staff.'[49]

The harmful impacts of capitalist productivism not just on workers' health but also on the surrounding communities had, then, already been acknowledged:

> Every month, for every square kilometre of Mestre, 10,000 kilos of powder produced in the Porto Marghera plants rain down. Obviously, the powder comes with gases whose toxic effects are universally recognised. Lung cancer rates in the Mestre and Venice population are among the highest in Italy.[50]

Following Panzieri, a critique of capitalist technology was also in the equation.[51] Safer machinery was necessary, 'but maybe a new breed of engineers is required to build machines that would not destroy health and enormously increase profits'.[52] The group, in fact, criticised the union line of demanding more growth rather than shorter hours to maintain employment levels, as such investments would result in automation and layoffs, and 'the bosses are still using automation in an anti-worker function. The introduction of new machines and computers does not lead to a decrease in working hours but to an increase in their profits.'[53]

The group also intervened on the noxiousness faced by outsourced workers:

48 Tronti, *Workers and Capital*, 358–87.

49 Potere Operaio flyer, 21 September 1968.

50 Potere Operaio flyer, 28 November 1968.

51 Raniero Panzieri, 'The Capitalist Use of Machinery: Marx Versus the "Objectivists"' [1961], in Phil Slater (ed.), *Outlines of a Critique of Technology*, Atlantic Highlands, NJ: Humanities Press, 1980, 39–68.

52 Potere Operaio flyer, 28 November 1968.

53 Comitato Operaio flyer, 6 February 1970.

> Outsourced workers are exposed more than others to accidents, often deadly, or to the nastiest diseases caused by gases, powders, mercury vapour, solvents, acid substances, etc. Comrades, it is true that all workers are exploited, but the outsourced workers are exploited twice [by the subcontracted enterprise and by the main company].[54]

These conditions were so unbearable that a three-month strike by the outsourced workers turned into a riot.[55] From 3 to 5 August 1970, the workers – including many residents of the Ca' Emiliani neighbourhood – barricaded the streets of Marghera. The most violent clashes took place near the Church of Jesus the Worker, where the police fired on the demonstrators before an ameliorative agreement was eventually reached. In these actions, the most relevant organisation was the 'rival' extra-parliamentary group Lotta Continua (Continuous Struggle), also active in struggles against noxiousness.

The Porto Marghera *operaista* group's reflections were systematised in two documents: 'The Refusal of Work' manifesto and the paper 'Against Noxiousness'.[56] The latter began by stating that:

> It is necessary to immediately distinguish between one form of noxiousness – i.e., as it is traditionally understood – linked to the work environment (toxic substances, smokes, powders, noise, etc.), from the one linked more generally to the capitalist organisation of work. No doubt, ultimately, the second type of noxiousness has a deeper impact on the worker's psychophysical balance. It makes him [*sic*] an alienated entity, a piece of the productivist machine completely detached from the end of his work, subject to the continuous usury that brings upon him an inhuman use of his labour

54 Comitato Operaio flyer, 3 July 1969.

55 Michele Boato, *La lotta continua*, Venice: Libri di Gaia, 2019.

56 Workers Committee of Porto Marghera, 'The Refusal of Work' [1970], *LibCom*, 11 October 2012, libcom.org; Political Committee of the Porto Marghera Workers, 'Against Noxiousness' [1971], *Viewpoint Magazine*, 1 April 2021, viewpointmag.com.

> power as it is the capitalist one, driven exclusively by the profits of the ruling class.[57]

The document went on to note that, while in former capitalist phases traditional noxiousness was necessary to keep costs low, its reduction had become compatible with the advanced capitalism of the time. On the one hand, traditional noxiousness mitigation provided a political justification for technological changes that downsized the workforce; on the other hand, it was inefficient to wear down too quickly the educated workforce required to operate the machines. Parts of the text are eerily prescient: 'In the new factory, coupled with a modest reduction in toxicities and thus in occupational diseases traditionally understood, there will be a strong increase in mental health disorders'.[58] A struggle for health that targeted only traditional noxiousness was deemed insufficient because it would be harnessed to the requirements of capitalist restructuring while leaving the crux of the matter – the prioritisation of value production over life reproduction – untouched.

'Against Noxiousness' also identified working-class communities as sites of class struggle at the point of reproduction: 'The working-class neighbourhood . . . is a big cage where the proletarians are locked up to squeeze a bit more out of them . . . A blatant environmental noxiousness is to be found there as a result of smoke pollution from the factories.'[59] This was in line with the feminist analyses developed at the time by Mariarosa Dalla Costa and Alisa Del Re around working-class reproduction.[60]

On 2 November 1972, the Porto Marghera group launched the Assemblea Autonoma di Porto Marghera (Autonomous Assembly of Porto Marghera) in an attempt to reach out to workers who did not

57 Political Committee, 'Against Noxiousness'.

58 Ibid.

59 Ibid.

60 Dalla Costa and James, *The Power of Women*; Lucia Chisté, Alisa Del Re and Edvidge Forti, *Oltre il lavoro domestico: Il lavoro delle donne tra produzione e riproduzione*, Milan: Feltrinelli, 1979.

identify with Potere Operaio. The latter had become increasingly riven by internecine struggles, which resulted in both the Porto Marghera group and the group around Negri (who had moved his political activities to Milan) leaving the organisation after the May–June 1973 congress. Potere Operaio would cease to exist by the end of the year. As the subsequent efforts to coordinate the different factory assemblies at a national or regional level failed, the Porto Marghera group's activities remained focused on a mostly local level. This was signalled by the founding of the bulletin *Lavoro zero: Bollettino della Assemblea Autonoma di Porto Marghera*, directed by Augusto Finzi, with the first issue released in October 1973.

The Porto Marghera group's theory of noxiousness developed between 1967 and 1972, chiefly by employees of polluting factories with no university education, was in many ways ahead of its time in locating health damage and environmental degradation as terrains of struggle intrinsically linked to capitalism. And yet, the proposal to address noxiousness they put forward in those early years appears somewhat blunt. In fact, the focus on 'more money, less work', and the list of quantitative demands resulting from it, did not directly tackle noxiousness itself (in principle, an industry can be toxic even if it employs zero workers) and appeared to subscribe to the notion that an imminent collapse of capitalism could be provoked through a factory-centred strategy of refusal. The group's subsequent reflections, however, led them to further extend their interventions in the community and to elaborate a new brand of qualitative demands.

The connection between factory and community struggles had been experimented with since at least 1970, but it reached a high point in 1974. With the 1973 'oil crisis', which inaugurated Petrolchimico's slow decline, the gains that had been made by workplace struggles in the sphere of production were being eroded by inflation in the sphere of circulation. In response, Assemblea Autonoma – in coordination with neighbourhood committees and feminist collectives – organised a campaign of price 'self-reductions' for utility bills, transport fares and basic consumer goods, including pickets at supermarkets and house occupations. In December 1974, 13,000 electricity bills were self-reduced and an agreement between unions and government led to decreased utility prices.[61]

Noxiousness came to the fore once again after the new Petrolchimico toluene diisocyanate (TDI) units were completed in 1971 and repeated leaks of highly toxic phosgene and other gases poisoned hundreds of workers.[62] On 23 November 1973, an Italian military aeroplane crashed on Petrolchimico's parking lot, not far from the phosgene tank. A catastrophe was averted by sheer luck. In January 1973, the Labour Inspection required all Porto Marghera employees to work with a gas mask within reach. The order caused widespread outrage, not only due to its unworkability, but also because it implied that perilous working conditions need not be eliminated.

It was in this context that Assemblea Autonoma organised the initiative at the factory gates that staged the crucifixion of a mannequin in a gas mask. The action was publicised with a flyer stating:

> Workers do not enter factories to make inquiries, but because they are forced to do so. Work is not a lifestyle, it is the obligation to sell oneself to make a living. It is by struggling against work, against the coerced sale of themselves, that workers clash against all norms of society. It is by struggling to work less, not to die poisoned by work, that they also struggle against noxiousness. Because it is noxious to wake up every morning to go to work, it is noxious to accept productive paces and conditions, it is noxious to take the shift system, it is noxious to go home with a wage that forces you back into the factory the day after.[63]

Noxiousness also loomed large in the units handling vinyl chloride monomer (VCM), an intermediate material for PVC production. Porto Marghera's worker-poet Ferruccio Brugnaro wrote: 'Vinyl chloride/

61 Sacchetto and Sbrogiò, *Quando il potere è operaio*, 82.
62 Bortolozzo, *L'erba ha voglia di vita*, 145–7.
63 Assemblea Autonoma flyer, 26 February 1973.

spares no one/ death was/ never/ so present.'[64] In 1974, Assemblea Autonoma published a pamphlet informing the workers on the most dangerous substances used in the industrial complex, including vinyl chloride, with a detailed analysis of their presence in the labour process. The group rejected the jobs versus environment dilemma, demanding that the most noxious units be closed down: 'Just like the monetisation of health is a losing game, so is the defence of employment at any cost . . . [it] risks becoming a new form of monetisation, involving the negotiation of the number of workers to be hit by cancer every year.'[65]

Assemblea Autonoma's last publication was a pamphlet titled 'Absenteeism: A Terrain of Workers' Struggle'. The essay recognised that the healthcare reform of the time was a concession to workers' struggles against noxiousness, while charging that such struggles were being harnessed to meet capitalist requirements for an educated and durable workforce: 'The protection of health is the planned wearing down of the workforce . . . From the planned obsolescence of the means of production and of commodities, we arrived at the planning [of the obsolescence] of the capacity to work.'[66] According to the group, such a rationalisation would bring the unions into co-managing the repression of large-scale absenteeism, 'one of the forms in which the workers' foreignness to the mode of production is expressed'.[67] The pamphlet dismissed triumphalist interpretations of mass absenteeism, acknowledging its individualism and poor amenability to collective organisation, but arguing provocatively that it was a legitimate form of struggle among others.

In early 1975, Assemblea Autonoma dissolved due to internal divisions and its failure to become an organisational alternative to the unions. The Porto Marghera group thus transformed *Lavoro zero* from a

64 Ferruccio Brugnaro, 'Il cloruro di vinile', in *Mortedison: Tutti assolti*, Milan: Tam Tam, 2001, 11; see also Ferruccio Brugnaro, *Fist of Sun*, Willimantic, CT: Curbstone Press, 1998; RoseAnna Mueller, 'Ferruccio Brugnaro: Italy's Proletarian Poet', *Italica* 92(3), 2015, 691–701. Ironically, Ferruccio Brugnaro is the father of Venice's incumbent conservative mayor Luigi Brugnaro, a businessman who rose to prominence thanks to his temp-work agency.

65 Assemblea Autonoma pamphlet, 1974.

66 Assemblea Autonoma pamphlet, 1975.

67 Ibid.

mimeographed factory bulletin into a printed magazine titled *Lavoro zero: Giornale comunista dal Veneto* (LZ), once again directed by Augusto Finzi, but now more professionally designed and accompanied with countercultural graphics. From 1977, *Lavoro zero* was supplemented by the weekly paper *Controlavoro* (CL).[68] Between 1975 and 1980, the Porto Marghera group remained politically active in the factories and communities, but its reach declined as – in the context of economic restructuring – the epicentre of radical mobilisation shifted from the large factories to the universities and young precarious workers. The Porto Marghera group did not join the then emerging autonomist organisation Collettivi Politici Veneti (Veneto Political Collectives). However, around 1979, some members would part ways with the original group to found a new Comitato Operaio (Workers' Committee) linked to Collettivi Politici Veneti.

In this period, Augusto Finzi, Gianni Sbrogiò and others (including activists from different groups, such as Lotta Continua's Michele Boato) launched Collettivo di Lotta contro le Produzioni Nocive (Collective of Struggle against Noxious Productions, CLPN). This was a rather extraordinary initiative as it inverted the typical (albeit not universal) scenario of polluting industry workers defending their industries against infuriated fenceline communities. Here, the factory workers themselves published data on industrial emissions and waste disposal, conducted participative inquiries on noxiousness not just in factories but also in communities, and organised public meetings and displays to sensitise the local population to the health and environmental harms caused by their own employers. For example, a CLPN statement directed to the fenceline communities read:

> Every year, 1,200 vinyl chloride tonnes are emitted by Montedison's chimneys . . . We are all Montedison employees. The struggle against noxiousness is a common struggle, health protection does not concern those who work inside the factories only . . . We are organising an

68 All issues of *Lavoro zero* and *Controlavoro* can be found in the online archive leftove.rs.

> inquiry to verify the health damage caused by vinyl chloride on the population of Marghera and Mestre's communities.[69]

While the CLPN's campaign did not reach mass proportions in the communities, their inquiries represented an interesting inversion of the *operaista* co-research method. In the 1960s, intellectuals external to the factories encouraged workers to research their abode of production to co-produce knowledge aimed at political action. With the CLPN, it was factory workers who encouraged external subjects – mainly women and youth – to research their abode of reproduction: 'It is necessary that women in the neighbourhoods manage this initiative directly, assessing it and implementing it if they think it is useful.'[70]

The influence here of the feminist theories discussed above is highly probable. In fact, while the Porto Marghera group's make-up mostly reflected the male-dominated composition of the area's industrial workforce, *Lavoro zero* hosted contributions from feminist collectives active in nearby communities. For example, in 1975, an article linked to the Wages for Housework campaign highlighted the importance of community-based struggles, indicating the need

> to break to our advantage the ghettos (family, neighbourhood, parish, village) through which capital has historically ruled society . . . We can interpret along these lines the occupations of luxury houses, the demand for schools at the service of the communities, the demand for control on environment and health [against] noxiousness, and the demand for wages for housework.[71]

While reaching out to communities, the group concluded that it was necessary to also demand deep qualitative transformations of production:

69 CL, 30 May 1977, 4.
70 CL, 30 June 1977, 3.
71 LZ, December 1975, 28.

> Today we need to challenge the organisation of work, not just its levels and parameters, but also the fundamental choices that make us produce commodities and not wealth, useless things made only to generate profits . . . *What, how and how much to produce* must be our parameters of struggle.[72]

For example, as a detailed information sheet on VCM-PVC noxiousness stated:

> Workers and proletarians need to decide on productive choices based on our real collective needs . . . As PVC is not biodegradable, the result is an enormous mass of waste that could only be eliminated with out-and-out recycling factories . . . We have good reasons to believe that an increase in PVC production is not consistent at all with our needs.[73]

This qualitative discussion was linked to the concept of 'proletarian self-valorisation', meaning working-class empowerment through the production, reappropriation and sharing of use values for the satisfaction of collective needs, as opposed to capitalist work for the production of exchange values: 'It is a struggle . . . between use values invented through creativity and the scrap, the so-called commodities, produced by coerced labour.'[74] The group connected self-valorisation to the struggle against noxiousness, proposing 'a struggle for the self-management of health and the full democratisation and socialisation of instruments and services, demanding social and health measures that improve the work environment as well as the environment of collective life'.[75]

Demands for a qualitative transformation of production went hand in hand with a redoubled repudiation of the neutrality of technological

72 CL, 19 March 1979, 1.
73 CL, 13 June 1977, 2–3.
74 LZ, December 1977, 4.
75 LZ, May 1979, 15.

progress, extending to a critique of science to the extent that it is inserted into the valorisation of capital. The May 1979 *Lavoro zero* issue dedicated to science appears to echo Augusto Finzi's reflexivity on his role as a factory technician, a frontline expert. *Lavoro zero* did not contest the reality of scientific discoveries, but noted how expertise is subordinated to the imperatives of capitalist accumulation:

> The new particles are good for advertising, but the real motive [of research] lies in what most scientists – in their ideological perversion – see as by-products or as unworthy of consideration: high-speed electronics, complex data elaboration systems, superconductors, interpretative models, materials technology, the organisation of work, the possibility of making arms . . .[76]

Technology developed under capitalism is not a set of tools that can be unproblematically repurposed to further workers' autonomy. Indeed, 'the proletariat cannot compete with capital on the level of weaponry, nor wishes to do so because such weapons bear in their DNA the conditions to create monopolies of violence'.[77]

This antagonistic-transformative approach to technology, and the associated demands for deep qualitative changes in production, were deployed in various noxiousness-related debates, such as those concerning food, medicine and energy. Regarding the latter, *Lavoro zero* criticised nuclear energy as a further step in the centralisation, verticalisation and militarisation of energy production, allowing capital to strengthen its 'energy blackmail' of the working class. In response to this blackmail – in the form of austerity and restructuring following the 'oil crisis' of the 1970s – solar energy was proposed as a way forward. Not only for its lower environmental impact, but also for its amenability to political decentralisation: 'We are not prefiguring a society fragmented into isolated producers or a return to craftmanship . . . The object of

76 Ibid., 33.
77 Ibid., 39.

decentralisation is not the proletariat but the current capitalist configuration.'[78]

In the course of the 1970s, then, the quantitative demand for 'more money, less work', proposed by early *operaismo* in its strategy of refusal, turned out to be insufficient. Worktime reductions might spare workers from workplace-based noxiousness but they would not necessarily halt environmental degradation. This problem called for a qualitative struggle over 'what, how and how much to produce', striving towards a sustainable relationship between human and non-human nature. In other words, it was deemed 'more valid to change the content of production rather than taking over the productive apparatus [as it is]'.[79] *Controlavoro* reflected on this shift in its penultimate issue: 'The contradiction existing within every one of us between a "desire to produce" and the "refusal of work" pushed us to attempt "productive" experiments that were qualitatively and quantitatively different from the production we endured every day.'[80]

In many ways, the Porto Marghera *operaista* group self-consciously reflected Veneto's regional specificities. And, yet, its trajectory was also inextricably linked to that of Italy's Long 1968 and similarly moved from the enthusiastic discovery of workers' autonomy to a dark ending. On 21 December 1979, Finzi, who had quit his Montedison job the previous year to pursue other projects around alternative medicine and communication, was arrested along with others as part of the '7 April 1979' prosecution against autonomist intellectuals and organisers. More were arrested on 24 January 1980, including Gianni Sbrogiò. While most would soon be released, Finzi spent two years and ten months in prison and Gianni Sbrogiò four years. The *Lavoro zero* issue published in April 1980 was thus the last of fifteen. It was almost entirely dedicated to the Minamata disease, discovered as a result of mercury contamination at a Japanese chloralkali plant. The experience of the group closed with the warning that the struggle against noxiousness could not be postponed to

78 LZ, December 1976, 14.
79 CL, January 1980, 1.
80 CL, 15 July 1980, 1.

'after the revolution' as traditional Marxism was deemed to believe.[81] This was not, however, the only subject of the issue.

On 29 January 1980, Sergio Gori, Petrolchimico's deputy director, was assassinated by the Red Brigades. This was followed, on 12 May 1980, by the murder of Mestre's police deputy commissioner Alfredo Albanese. Finally, in 1981, the Red Brigades kidnapped and killed Giuseppe Taliercio, the Petrolchimico director. The Porto Marghera group condemned the Red Brigades' actions:

> The terrorists can attack today, in a context [Veneto] where they had never had political space, because the powers that be clear the way for them, liquidating forms of organisation and struggle that move in a totally different direction . . . For what concerns, for example, the dreadful price that Marghera's industrial plants impose on the workers and the population in terms of noxiousness and quality of life, our research was always based on the attempt to indicate mass alternatives on how, how much, and what to produce.[82]

However, the Red Brigades' senseless assassinations effectively annihilated the group's activism in Porto Marghera's factories. In fact, they provided a justification for heavy state repression and delegitimated the Porto Marghera group, engendering a climate of suspicion and provoking the charge that their inflammatory rhetoric had contributed to the killings.

In 1981, Montedison announced the layoff of 8,000 workers, 616 of whom were in Porto Marghera. The mobilisations to defend the jobs were to no avail. Like the 1980 defeat of FIAT autoworkers in Turin, this blow marked the breakdown of workers' intransigence towards restructuring, ushering in the neoliberal phase in Italy. Meanwhile, a new form of noxiousness, heroin addiction, infiltrated working-class neighbourhoods across countries, taking a high toll in Marghera too.

81 LZ, April 1980, 3–4.

82 Ibid., 2.

After serving his prison time, Finzi founded an environmentalist association but abandoned political militancy. In 2002, he reflected:

> My removal [as a reprisal for the 1968 strikes] from the infamous CV6 [vinyl chloride] unit maybe ended up saving my life, because almost all those who worked there are dead by now. I was effectively put in an 'exile unit', in the infirmary . . . Here, I discovered a new figure that I had never considered before, that of the sick person. One thing is seeing the worker as a subject of antagonism, the moment of struggle, of rage . . . Another thing is seeing him [*sic*] as a person who suffers . . . Health seen as a tension lived inside, not just damage caused by production.[83]

Finzi died of brain cancer in 2004, aged sixty-three.

The Porto Marghera group's approach, despite the limitations and mistakes that can be attributed to it, represents an early, inventive case of working-class environmentalism that contributed to the winning of substantial concessions. Their critique of capitalist work and technology, the connections they made between workplace and community struggles, and their articulation of both quantitative demands (wealth redistribution and short working hours) and qualitative ones (the transformation of production in an ecological direction) are still useful for today's attempts to break the link between workers' reproduction and capitalist noxiousness. However, subsequent changes in working-class composition have weakened the possibility of a mass diffusion of such radical platforms in capital-intensive workplaces. Their schema, therefore, cannot be slavishly replicated in our warming days.

Marghera vs Marghera, Italy

Marghera, the neighbourhood, is the Other of Porto Marghera, the industrial area. Here, the separation between production and reproduction – with its structurally uneven distribution of activities in space – is

83 Interview with Augusto Finzi by Manuela Pellarin, 2002.

at first glance closely mirrored by physical geography: the two Margheras are divided by a single road, Via Fratelli Bandiera, severing the dark smokestacks from the red-tiled condos. In 2000, the Venice council area had a population of 275,000 inhabitants, 28,000 of them living in Marghera. Most people in the Veneto region know the latter indirectly, via a mythology of immense and deadly factories, endemic heroin addiction, sex trafficking and crime, as well as left-wing 'extremism' yesterday and 'excessive' immigration today. To some extent, this picture clashes with that offered by the neighbourhood when one actually visits it: large and well-kept green areas, strong public services, lively and crowded open spaces.

The province of Venice, however, has consistently reported deaths and cancers beyond the regional average.[84] Recent research has also demonstrated that, within the province, Marghera has the highest excess deaths and cancer incidences, although the difference between Marghera and other areas has been narrowing in recent years.[85] Unsurprisingly, Marghera's poorest neighbourhood, the slum of Ca' Emiliani, was the nearest to Petrolchimico (before being flooded and mostly demolished in 1974). Indeed, a medical report from 1971 stated: 'The urban organisation of Mestre-Marghera means that the lowest-income people live where the air is most unhealthy.'[86]

Over the decades, Marghera saw two competing working-class environmentalist perspectives emerge from the 1960s and '70s struggles against noxiousness, one mostly organised in the factories and the other in the community. Both based their proposals on the principle that 'social' and 'environmental' rights should be mutually reinforcing rather than opposed, and their appeals to the working class were made credible

84 Amerigo Zona et al. (eds), 'Sentieri: Sesto rapporto', *Epidemiologia e prevenzione* 47(1–2), Supplement 1, 2019.

85 Francesca Gessoni, Silvia Macciò, Claudio Barbiellini Amidei and Lorenzo Simonato, 'Study on the Health Status of the Population Living in Marghera (Venice, Italy) through the Use of a Longitudinal Surveillance System', *Annali dell'Istituto Superiore di Sanità* 56(2), 2020, 157–67.

86 Cited in Gianfranco Bettin (ed.), *Petrolkimiko: Le voci e le storie di un crimine di pace*, Milan: Baldini & Castoldi, 1998, 40.

by their sizable working-class constituencies, among industrial workers for the workplace-centred camp and among fenceline residents for the community-centred camp. Marghera, with its rough and rowdy reputation, does indeed feature a lower average income than the council area as a whole (a proxy for high working-class presence).

The workplace-centred camp was composed of the 'confederal' unions, the largest being CGIL, and the majorities within the centre-left parties.[87] The community-centred camp had also originated in the workplace with the New Left factory groups, particularly the Porto Marghera *operaista* group. However, after being marginalised in the factories, the New Left put down stronger roots in the neighbourhoods. The community-centred camp's composition shifted to comprise environmentalist and workers' health associations grounded in the history of struggles against factory noxiousness (most prominently Ambiente Venezia, Associazione Bortolozzo, Ecoistituto Langer and Medicina Democratica), the leftist Partito della Rifondazione Comunista (Communist Refoundation Party, PRC), the Green Party, the post-autonomist Rivolta Social Centre and the radical union Confederazione Unitaria di Base (Unitary Rank-and-File Confederation, CUB).

In traditionally left-wing Marghera, opposition to industrial pollution contributed to the popularity of the New Left, as shown by the fact that all borough presidents between 2005 and 2020 belonged to PRC or the Green Party. Many Green Party local leaders, such as Gianfranco Bettin and Michele Boato, had formerly belonged to the extra-parliamentary group Lotta Continua. The local CUB union branches and the Rivolta Social Centre, by contrast, emerged mainly from *operaismo* and autonomism. In the early 2000s, CUB still represented a vocal minority of Porto Marghera workers, and within Petrolchimico, which had gradually been bought up by Ente Nazionale Idrocarburi (National Hydrocarbon Agency, ENI), CUB was led by environmentalist

87 In the Italian labour jargon, the confederal unions are the major labour confederations. The phrase is used to distinguish them from the smaller and radical 'Cobas' unions, whose origins are in workers' rank-and-file committees from the late 1960s.

employees such as Luciano Mazzolin.[88] The Rivolta Social Centre is based in a large, disused factory in Porto Marghera. It was self-managed by the local branch of the post-autonomist Tute Bianche (White Overalls) movement, becoming an important social milieu for working-class youths (for example, Venice football fans and the Venezia Hardcore music collective).

The workplace-centred and community-centred camps were both represented in the Venice administrations of the 1990s and 2000s, mostly led by 'mayor-philosopher' and former *operaista* leader Massimo Cacciari. Even Tute Bianche had their activist Beppe Caccia elected as a city councillor with the Green Party. A significant point of contention between the two camps opened up, however, over Petrolchimico's so-called 'chlorine cycle': three integrated plants located downstream from 'Petrolchimico's heart', its ethylene cracker, producing chlorine and sodium hydroxide, PVC and TDI respectively.

The community-centred camp held that Porto Marghera's chlorine cycle constituted an unacceptable source of risk and health damage for the densely populated area. The three ageing plants were subject to frequent leaks and accidents, alongside toxic emissions, the contamination of soil and water, and the storage of hazardous substances such as chlorine, vinyl chloride and phosgene. The community view was the plants had to be closed with environmental remediation needed to make space for cleaner production. Their position on employment was that workers' income had to be protected through a combination of subsidies, employment in remediation works and redeployment in other local factories (already existing or to be created post-remediation). Additionally, they supported Greenpeace's calls for a global decrease in plastic production (particularly PVC).

The workplace-centred camp agreed that Petrolchimico had been a major noxiousness supplier in the past but held that the trajectory of substantial technical improvements since the 1970s could progress

88 ENI, Italy's main oil and gas company, operated in Porto Marghera through a variety of holdings and branches (Enichem, Polimeri Europa, Syndial, Versalis, etc.). Here they are all referred to as ENI for simplicity's sake.

further, achieving a sustainable cohabitation between the chlorine cycle and communities. With the double objective of relaunching industrial employment in Porto Marghera while improving the industries' environmental performance, the Agreement for Chemistry between trade unions, employers (mainly ENI) and public authorities was signed on 21 October 1998.

Many activists in the community-centred camp recognised the heritage of Porto Marghera's 1970s workplace-centred environmentalism but held that changing employment patterns had driven a wedge between factories and neighbourhood. As janitor and Rivolta leader Roberto Trevisan told me: 'My partner's dad used to work with vinyl chloride, and he died of lung cancer . . . Almost everyone [from Marghera] in my generation, I'm sixty, had someone in the family who died or had health issues . . . Then employment went down because, obviously, technological innovation lessened the demand for labour. The wall they built along Via Fratelli Bandiera is also a political wall, separating the factories from the city.' In turn, Gianfranco Bettin, born and raised on a Ca' Emiliani council estate, recalled: 'In Marghera a different environmentalism emerged, in a narrower version even from within the factories . . . Some trade unionists did try to say we were elitist like previous environmentalists. But it was frankly untenable because some of us worked in the factories, or were the children, siblings, wives, or partners of factory workers. I mean, we were part of the working class.'

Just as the community-centred camp rejected the 'elitist' label, the workplace-centred camp insisted that it too cared for the environment. Petrolchimico union activist Nicoletta Zago reflected: 'The lab was my life, it was a job I always liked. This is why my pride was on the line with this story [the possible plant closure] and I've always considered it to be an injustice . . . In any case, there had been a process of "environmentalisation" at Petrolchimico, that is, caring more for safety, workers' safety and environmental safety . . . We too were part of environmentalism.'

However, the relationship between the two camps had always been strained. The basis of this divergence arguably lay in the discrepancy between their social and technical compositions. Throughout its history, only a minority of the industrial area's workers lived in Marghera, which is hardly surprising as for many decades the number of jobs in Porto Marghera exceeded that of Marghera residents. However, in the 1960s, when Porto Marghera employed between 33,000 and 40,000 workers, a high share of Marghera residents were employed in the factories.[89] This gradually changed when employment in the industrial area started to decline in the 1970s (see Figure 4.2).

Figure 4.2 Number of employed in Porto Marghera

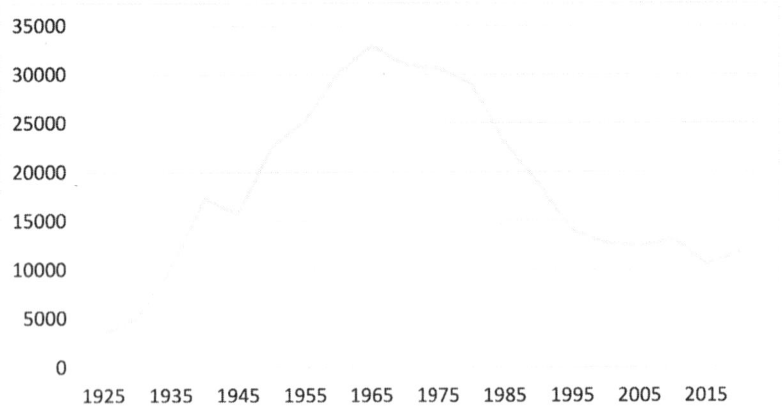

35000
30000
25000
20000
15000
10000
5000
0

1925 1935 1945 1955 1965 1975 1985 1995 2005 2015

Source: Ente della Zona Industriale di Porto Marghera.

By the 1980s, the relationship between neighbourhood and factories had become increasingly uneasy as the locals found fewer and fewer jobs on the other side of Via Fratelli Bandiera. This malaise surfaced provocatively when the local reggae band Pitura Freska rose to fame singing: 'Marghera without factories would be healthier: a jungle of corn, tomatoes, and marijuana' ('Marghera', 1991). Marghera was undergoing a deep noxious employment deindustrialisation process, with Petrolchimico's slow-motion downswing symbolising the crisis of the whole area.

89 The difference in the figures depends on whether one includes an estimate of the workers under the most transient employment arrangements. Valentina Bonello, 'Porto delle nebbie e luoghi chimici: Marghera, etnografia della città post-industriale', PhD thesis, University of Verona, 2017, 179.

The chlorine cycle directly employed about 600 workers, a large majority of whom commuted from the outer urban centres of Mestre and Chioggia, as well as rural localities scattered within and beyond the province. In turn, many Marghera residents commuted to insular Venice to work in its public services, hotels, shops, restaurants and cultural sector. Based on the framework adopted here, the majority of Marghera residents were working class. However, as they were less industrially employed than before, most did not perceive the possible closure of the chlorine complex as a threat. This widening divergence between community and workplace set the stage for a political bifurcation between those employed in the chlorine cycle and those working elsewhere but living close to it.

In the summer of 1994, retired petrochemical worker Gabriele Bortolozzo entered the office of Public Prosecutor Felice Casson to submit a complaint against his previous employers Montedison and ENI, based on his own lay research on the health of his colleagues at Petrolchimico. The blue-collar worker turned green activist believed that the high number of cancer deaths was due to exposure to vinyl chloride.[90] Casson did open an inquiry, but Bortolozzo never saw the beginning of the trial. He was killed in a road accident while riding his bicycle on 12 September 1995.

Large numbers turned up for the first hearing of the Petrolchimico case on 13 March 1998, taking place in the Mestre courtroom built in the late 1970s as a bunker to protect it from armed attacks.[91] Thirty Montedison and ENI managers and doctors stood accused of charges including manslaughter, injury and environmental disaster, while hundreds of former Petrolchimico workers, relatives of the victims, unions, NGOs and public authorities participated as civil parties. The charge of manslaughter

90 Gabriele Bortolozzo, 'Storia di un killer chimico: Il cloruro di vinile', *Medicina democratica*, 92/93, 1994, 32–92.

91 Barbara L. Allen, 'A Tale of Two Lawsuits: Making Policy-Relevant Environmental Health Knowledge in Italian and US Chemical Regions', in Christopher C. Sellers and Joseph Melling (eds), *Dangerous Trade: Histories of Industrial Hazard Across a Globalizing World*, Philadelphia: Temple University Press, 2011, 154–67.

concerned the death by cancer of 149 Petrolchimico workers who had been exposed to vinyl chloride, although subsequent research updated the number to 248.[92] The charge of environmental disaster referred to soil and water contamination, the severe pollution of the Venice Lagoon, and the proliferation of noxious landfills.

The stakes were high, as the legal case had become a battle over the narrative of Porto Marghera's past with implications for its future.[93] The workplace-centred and community-centred camps both endorsed the trial, but they diverged on how to interpret it. According to the community-centred camp, the lawsuit contributed to their contention that the chlorine cycle plants should be closed. The workplace-centred camp, led by CGIL's chemical federation local leader, Bruno Filippini, countered that union action had substantially reformed Petrolchimico, and thus – while the trial was needed to address past wrongdoings – it had no implications for the future. These differences reflected an intra-working-class split that saw, on the one side, most workers currently employed in the chlorine cycle and, on the other side, a range of service sector and reproductive workers living in the area, with retired industrial workers and middle-class allies more evenly distributed between the two camps.

CUB was the only workplace-based organisation backing the community-centred camp. As the most direct inheritor of 1970s' *operaista* environmentalism, the radical union was adamant about holding on to the connection between workplace and community struggles. For example, in 1998 and 1999, prosecutor Luca Ramacci requisitioned the main Petrolchimico wastewater treatment plant and its SG31 incinerator under the charge that its waste disposal practices violated legal requirements (15 ENI managers would plead guilty two years later). In response, the confederal unions staged strikes and roadblocks, but CUB criticised their move: 'The workers must indeed defend their right to employment but without becoming hostage to those who only care for profits, and

92 Roberta Pirastu, Michela Baccini, Annibale Biggeri and Pietro Comba, 'Epidemiologic Study of Workers Exposed to Vinyl Chloride in Porto Marghera: Mortality Update', *Epidemiologia e prevenzione* 27(3), 2003, 161–72.

93 Laura Cerasi, *Perdonare Marghera: La città del lavoro nella memoria post-industriale*, Milan: FrancoAngeli, 2007.

without agreeing to pay the price of diseases and deaths among workers and the population.'[94] However, since the early 1980s, the relentless haemorrhage of industrial jobs and resulting cogency of dismissal threats eroded the material basis of *operaista* sway in the factories. By the late 1990s, CUB was far from having enough support in Petrolchimico to pressure the confederal unions to adopt a different approach.

The first-instance ruling came on 2 November 2001, after 150 testimonies and 1.5 million pages of documentation. The bunker-courtroom was packed and the atmosphere tense. Judge Ivano Nelson Salvarani slowly read out his verdict, reiterating 'acquitted' for each and every charge. The public reacted with disbelief. Rivolta members led by Tute Bianche spokesman Luca Casarini broke into the courtroom and hung a banner reading 'Guilty, yesterday like today' under the inscription 'The law is equal for all'. A heated debate erupted across the media. What had happened?

The prosecution had shown that the carcinogenicity of vinyl chloride was first revealed in 1969–70 by Dr Pier Luigi Viola and confirmed to Montedison by Professor Cesare Maltoni in 1972.[95] Yet Montedison did not publicise the information due to a 'secrecy agreement' subscribed to by the largest petrochemical multinationals.[96] Carcinogenicity only became publicly known in 1974, after the correlation between vinyl chloride exposure and death from a rare cancer, called angiosarcoma, emerged in BFGoodrich's Kentucky plants. The court, however, rejected the secrecy agreement thesis, interpreting it as a confidentiality clause to be respected until research could be completed. Moreover, the judges accepted the defence's argument that, since 1974, Montedison had taken satisfactory safety measures.

Concerning the charge of environmental disaster, the court (like the defence) acknowledged the severe contamination of the soil, canals and lagoon by a great variety of pollutants (metals, polycyclic aromatic

94 CUB flyer, 22 June 1998.

95 Felice Casson, *La fabbrica dei veleni: Storie e segreti di Porto Marghera*, Milan: Sperling & Kupfer, 2007.

96 Gerald Markowitz and David Rosner, *Deceit and Denial: The Deadly Politics of Industrial Pollution*, Berkeley: University of California Press, 2002.

hydrocarbons, dioxins, etc.). Indeed, in December 1998, the government had declared Porto Marghera a highly polluted site in need of remediation. However, judge Salvarani held that it could not be proven that Montedison and ENI were the main culprits or had violated any existing laws. Ironically, two days before the verdict, Montedison had agreed with the Ministry of Environment (until then a civil party in the trial) a compensation of €271 million for environmental remediation.

The appeal court verdict was reached on 15 December 2004. Five defendants were convicted for the angiosarcoma death of Petrolchimico worker Tullio Faggian, because – as VCM's high toxicity had been known since the 1950s – Montedison had the legal responsibility of reducing workers' exposure even before carcinogenicity was fully proven. Judge Francesco Aliprandi thus recognised Montedison's responsibility for all other angiosarcoma deaths and Raynaud's syndrome cases, but the statute of limitations for them had expired. Additionally, the verdict acknowledged Montedison's responsibility for illegal discharges into the lagoon until the 1990s, but the statute of limitations for these irregularities had also elapsed. On 19 May 2006, the Cassation Court upheld the appeal court ruling.

Petrolchimico's troubles were not over, however. On 28 November 2002, a tank exploded in its TDI units. The sirens screeched through Marghera and the population was told to go home and shut their windows. For long minutes, the apocalyptic prophesy of a Marghera ablaze seemed to have at last come to pass. Luckily, however, the shockwave from a second exploding tank extinguished the main fire, allowing the firefighters to gain the upper hand. The fire department's official report later famously thanked all those 'who, at their own risk, prevented the accident from having worse consequences, certainly benefiting from the assistance of divine Providence'.

In what many saw as ENI's betrayal of the 1998 Agreement for Chemistry, Petrolchimico's TDI plant had been sold to Dow Chemical in 2001. In 2005, three Dow managers would be convicted for the 2002 explosion. However, after the first-instance court ruling in the Petrolchimico case, legal proceedings could scarcely appease the infuriated fenceline communities. This time, the focus of accusations was the deadly phosgene gas stored in a 15-tonne tank close to the TDI unit where the explosions happened.

On 4 December 2002, a large gathering of locals and activists led to the founding of Assemblea Permanente Contro il Rischio Chimico (Permanent Assembly Against Chemical Risk, APCRC). The APCRC united all the organisations that had already been campaigning against the chlorine cycle, as well as non-politicised residents whose trust in the industry had been broken by the accident. Even if some retired factory workers were still involved, the APCRC's most prominent leaders were a mixture of middle-class professionals, such as IT consultant Anthony Candiello and health and environmental inspector Franco Rigosi, and service sector workers, such as janitor Roberto Trevisan and social worker Michele Valentini.

Isabella Aprile, a catering worker and Green Party activist who grew up on a Marghera council estate as the daughter of a waste processing worker and a cook at Petrolchimico's canteen, reflected: 'In our group [of friends], all those who had lost their dads, their dads worked at Petrolchimico . . . So you figure out you don't want to end up like your dad and . . . Either you go away, and in fact most of them don't live in Marghera anymore, or you struggle to change the situation, and this is how you become an environmentalist . . . My mom fell ill when she was thirty-five and died at forty. Plus, you didn't need to work there [for your health to be affected], because Petrolchimico was here [close to her house].'

This exemplifies well what is meant here by community-centred working-class environmentalism, in which workers organise in their neighbourhoods to improve their conditions of reproduction.

Noting that 'jobs keep going down while cancers go up', APCRC began a long anti-chlorine campaign featuring several mass demonstrations.[97] Its largest event was the performance, on 23 November 2003, of

97 APCRC flyer, March 2003; Nicoletta Benatelli, Anthony Candiello and Gianni Favarato, *Laboratorio Marghera: Tra Venezia e il Nord Est*, Portogruaro: Nuovadimensione, 2006; Riccardo E. Chesta, Paolo Crivellari and Catalina Santana-Bucio, 'Histoire des mobilisations civiques sur les risques majeurs au Havre (France) et à Marghera (Italie)', *FonCSI Livrable* 2, 2014.

a play titled *Bhopal* by actor Marco Paolini, which attracted thousands to Marghera's market square. Paolini drew a parallel between the 1984 Bhopal disaster caused by an accident at Union Carbide's plant and the risks posed by the phosgene stored in Porto Marghera by Dow, which had in fact bought Union Carbide in 1999.

APCRC collected 12,600 signatures in support of a consultative referendum on the chlorine cycle in the council area. The run-up to the survey marked the highest point of acrimony between the two camps. Between 19 June and 8 July 2006, all Venice citizens received a postal ballot with the question: 'Do you want the production and processing of chlorine, vinyl chloride and phosgene to continue?' A week later, the results were out: 34.3 per cent of eligible participants had expressed their view and 80 per cent of these chose the 'No' option. Marghera was the borough with the highest participation rate (41 per cent), with 80 per cent against the chlorine cycle. The only area that delivered a 'Yes' majority was the rural neighbourhood of Trivignano, where a numerous group of Petrolchimico workers resided. If the Marghera result indicated that the community-centred camp had a significant working-class following, the Trivignano vote confirmed the same for the workplace-centred camp. However, Marghera residents voted according to their reproductive interests (a less hazardous and polluted environment), while the Trivignano residents voted according to their workplace-centred interests (job security).

In August 2006, Dow announced it would decommission its Porto Marghera plant. The corporation never revealed whether the survey had influenced its decision; Dow Italia manager Roberto Lombardi mentioned only 'market issues'. In any case, Dow's 180 workers were furious, as no 'green chemistry' investments were in sight for them to be redeployed. In December 2006, ENI agreed to re-hire the Dow workers who were not near retirement and would not find alternative employment, but this happened slowly and, for some, after years on income support schemes.

Meanwhile, remediation works were relaunched in 2002, using a combination of public funds and mandatory contributions by the industry, engendering hopes for a regeneration of the declining industrial area. However, between 2001 and 2005, INEOS acquired full ownership of Porto Marghera's PVC plants. In July 2008, INEOS announced it would abandon its Italian PVC plants, to the dismay of its 270 Porto Marghera employees. The factory was put under special administration, while the workers – led by technician Nicoletta Zago – began a tenacious campaign in defence of their jobs, including spectacular actions such as the symbolic occupation of the plant's flare stack and San Marco Square's bell tower. The remediation works that could have made space for new and cleaner jobs were progressing at a snail's pace and ENI was this time unwilling to re-hire the INEOS workforce, although extra-ordinary income protection schemes were kept in place until late 2014.

The PVC plants never resumed production. Eventually, in May 2022, ENI permanently turned off its ethylene cracker, to replace it with a plastic recycling plant and a waste incinerator whose construction was, however, successfully opposed by community groups. Despite the sudden and unplanned closure of the chlorine cycle, the bulk of Porto Marghera jobs were lost in the 1980s (see Figure 4.2). Since then, Marghera's employment composition has changed radically, with services rising and industry declining. The fall in house prices, combined with the availability of nearby employment opportunities, has in turn contributed to high immigration rates. Marghera thus shifted from a 2 per cent share of foreign residents in 2001 to 27 per cent in 2021, which mitigated the decline and ageing of its population.

An invisible part of this story was uncovered in 2013. It turned out that Consorzio Venezia Nuova (CVN) – the consortium of companies tasked with carrying out Porto Marghera's remediation – was behind a network of illicit financing (for an alleged €1 billion) and bribing of politicians in exchange for public tenders. Altero Matteoli – environment minister between 2001 and 2006 for the post-fascist party Alleanza Nazionale (National Alliance) – had secretly obtained, through a pact with CVN, the inclusion in Porto Marghera's remediation tender of entrepreneur Erasmo Cinque, a crony of his having no expertise to carry out such works. Matteoli made a lucrative deal after having brokered Montedison's €271 million environmental damage compensation in 2001. The tender for Porto Marghera's remediation – worth an ever-inflating sum of hundreds of millions – was then awarded to CVN without competition, and a portion of the finances ended up being embezzled through CVN's corruption network. During the CVN trial, businessman Piergiorgio Baita allegedly declared: 'Without CVN, Porto Marghera's clean-up would have been completed.'[98]

The environmentally sustainable chlorine cycle envisioned by the unions did not materialise, and neither did the regeneration based on post-remediation green industrial jobs advocated by the environmentalists. Moreover, the increased reliance of the local economy on tourism in Venice generated its own problems – and related social movements – with more precarious employment patterns, the further expulsion of Venetian residents to the mainland, the damage to the lagoon's ecosystem engendered by large cruise ships, and a considerable contribution to excessive air traffic.

And yet, the workplace-centred mobilisations did, to some extent, protect workers' incomes through the ENI re-hirings and the extraordinary income-support schemes. Likewise, the community-centred camp achieved significant environmental improvements in terms of partial remediation and reduced industrial pollution and hazards, while the Petrolchimico case ensured some compensation for the affected workers and their families. This was not the case in the UK, where David Foster – a VCM-exposed worker suffering from a liver condition – concluded: 'These people have got away with industrial murder . . . They've known about this bloody stuff for a long time. They fooled people into believing it was bloody harmless.'[99]

A convergence between the workplace-centred and community-centred camps could perhaps have more effectively pressured employers and public authorities to deliver environmental remediation that, to the present day, remains incomplete. Remediation would have been in the interest of both camps, as it would have removed much pollution from

98 Maurizio Dianese, '"Senza il Consorzio Marghera sarebbe stata tutta bonificata"', *Il Gazzettino*, 8 July 2016, ilgazzettino.it.

99 Cited in David Walker, 'Occupational Health and Safety in the British Chemical Industry, 1914–1974', PhD thesis, University of Strathclyde, 2007, 144.

the vicinities of residential Marghera, while making space for new sources of employment. However, even if both sides were sincere in their efforts to apply the principles of working-class environmentalism, the discrepancy between their social and technical compositions both obstructed the search for a unified political platform and obscured the common class base of their grievances. Nonetheless, as the ecological crisis deepens, the necessity for such a convergence between workplaces and communities is becoming more and more evident.

Conclusion
The Ecological Transition from Below

On 9 July 2021, the workers at Florence's GKN Driveline took over their car axle factory after an email announced that all of them, more than 400, had been fired with immediate effect. The former FIAT plant, located in the industrial area of Campi Bisenzio, was then owned by the British company Melrose Industries. The latter claimed that its decision had been made unavoidable by the shift to electric vehicles, as part of worthy efforts to ensure an ecological transition beyond fossil fuels. Shop steward Dario Salvetti, a leader of the GKN Factory Collective, commented: 'This narrative tells us that our layoffs are necessary to protect the environment: in *Il Sole 24 Ore* [the broadsheet of Confindustria, the association of Italian industrial entrepreneurs], a few days after we were laid off, an article came out which basically said: "You wanted Greta [Thunberg] and now you get layoffs."'[1]

While, in many such cases, workers and unions settle for negotiating enhanced redundancy benefits, sometimes holding a grudge against environmentalists, the GKN Factory Collective kickstarted a long struggle

1 Emanuele Leonardi and Mimmo Perrotta, 'Emanuele Leonardi and Mimmo Perrotta Interview Dario Salvetti', *Project PPPR*, 7 November 2022, projectpppr.org.

against the decommissioning.[2] Having witnessed and analysed so many past episodes of mass layoffs, they decided to play a more creative game. With the assistance of an interdisciplinary group of sympathetic researchers, they drafted a conversion plan to produce components for sustainable public transport and demanded its adoption via a nationalisation that would put the factory under workers' control.[3] On this basis, they proposed and sealed an alliance with the Italian climate justice movement and local environmental community groups.

On 26 March 2022, a demonstration jointly called by the GKN Factory Collective and Fridays for Future Italy took over the streets of Florence, drawing tens of thousands of people from all over the country. Blue collars and green activists, trade unions and environmental community groups, marched side by side. The personifications of the working-class-as-producer and of the working-class-as-reproduced, for once, recomposed. This strategy engendered a cycle of mass mobilisations, replicating this broad display of support multiple times. It was no longer just about these workers' jobs, but about contributing to a vision for a worker-led ecological transition. This placed considerable pressure on public authorities, so that the layoffs were twice overturned by Labour courts, while the bosses switched from their initial blitzkrieg to a long trench war aimed at wearing down the workers' determination.

However, neither Mario Draghi's 'technocratic' government nor Giorgia Meloni's 'post-fascist' one had any appetite to nationalise, let alone under workers' control, as that would have set a precedent for the many other factories under threat. The GKN Factory Collective thus decided to move forward with an attempt to establish a cooperative for the production of cargo bikes and solar panels. They explained their choice in these terms:

2 Francesca Gabbriellini and Paola Imperatore, 'An Eco-Revolution of the Working Class? What We Can Learn from the Former GKN Factory in Italy', *Berliner Gazette*, 17 April 2023, berlinergazette.de.

3 Collettivo GKN et al., *Un piano per il futuro della fabbrica di Firenze: Dall'ex GKN alla Fabbrica socialmente integrata*, Milan: Feltrinelli, 2022.

> One single factory can't be a model. But we are an example. Tomorrow, this example could generate statements, pamphlets, and analyses, but also concrete projects. For instance, the production of cargo bikes under workers' control, bikes that will be available to riders transforming the delivery industry, including its relationships to the urban environment. Photovoltaic panels at the service of energy communities, geared towards 'democratising' energy production and distribution. Local governments and public institutions could be part of the cooperatives and receive solar panels in return, and use them to eradicate energy poverty in their neighbourhoods.[4]

At the time of writing, the dispute is still ongoing, the permanent sit-in at the factory remains in place, and the popular shareholder campaign to kickstart production under workers' control is being relaunched (check out insorgiamo.org for all information on how to contribute). Since 9 July 2021, as the workers like to say, 'Not one bolt left this place.' From the inside, the factory offers an eerie sight: rows of machines standing still, boxes of finished axles under a thickening powder layer, as if time had stopped for years. Meanwhile, meetings in the refectory have gathered workers, trade unionists and activists from all around Europe and beyond (including Greta Thunberg, indeed), while the parking lot has become a venue for massive cultural events, from Oi! gigs to the Festival of Working-Class Literature. Time will tell whether this struggle will become a new episode in a long history of working-class heroic defeats (albeit one with an ecological slant), or, rather, an inspiring example for building alternatives to the jobs versus environment dilemma, or, perhaps, both.

Throughout this book, I have argued that the ecological crisis is rooted in the logic of capital accumulation. Noxious deindustrialisation – employment deindustrialisation combined with the persistence of industrial noxiousness – is one of its most salient manifestations, both

4 Luca Manes, '"Capitalism Is Anti-Us": Ex-GKN Workers Champion Ecological Transition' [2023], *People and Nature* (blog), 6 February 2024, peopleandnature.wordpress.com.

globally and in many local contexts. Noxious deindustrialisation was key to both the erosion of Fordist social contracts in the Global North and to a renewed increase in the surplus working class that has been particularly severe in the Global South. Such destructive tendencies can, however, be countered by a working-class environmentalism: one that begins from workers' reproductive interests as well as Global South workers' interest in refusing the current international division of labour and noxiousness, and aspires to become as diffused across the global working class as possible.

Capitalism is an inherently noxious mode of production because it subordinates the 'what, how and how much' of production to the impersonal forces of the market and the growth-bound profit imperative. The logical conclusion is that, in the absence of a global transition to an ecological form of socialism, it will be impossible to fully transform production to guarantee a sustainable metabolism between human and non-human nature. The trouble is, however, that such a global transition to eco-socialism is nowhere to be seen on the political horizon, and incendiary statements by radical academics or revolutionary verbalisms by left sects will not do much to resolve the impasse. Alas, humans make their own history, but not in conditions of their choosing.

In light of this, my suggestion is that the task of the moment is strengthening counterpowers for *working-class rigidity* against the flexibility required by the law of value, by weaving convergences between the struggles of different working-class segments. Or, as the GKN Factory Collective put it, by expanding the 'method of convergence'. Mainstream economists use the phrase 'wage rigidity' to designate the regrettable fact that workers are not prone to take wage cuts to ensure their employers' competitiveness. In the Italian labour movement, 'working-class rigidity' became a more general expression to indicate workers' refusal to shoulder the costs of crises, to bow to the requirement for pliancy that the market demands from them. Working-class rigidity is also known under the *nom de guerre* of 'dignity', something it is certainly worth fighting for.

A counterpower is only really so if it maintains the perspective of a systemic alternative. The 'green' plan of capital we encountered in Chapter 3 needed workers to be flexible (or 'smart', as the incumbent rhetoric would have it). Workers had to accept restructurings in the name of the ecological transition from above, thereby acquiescing to the logic of the system. Working-class rigidity, in contrast, is obtuse to such a logic. Yet it can sharply counterpose it with the alternative of an 'ecological transition from below'.[5]

Let me clarify what I do *not* understand the distinction between ecological transition from above and ecological transition from below to mean. Firstly, it does not refer to an empirical description of achieved processes, as we have not ecologically transitioned anywhere yet. It is rather a distinction between two different types of project, supported by different theoretical and discursive apparatuses. Secondly, it is not a Tocquevillian opposition between state and civil society. Thirdly, it is not an antinomy between 'verticalism' and 'horizontalism', or between a politics centred on social movements as opposed to one centred on institutionalised parties and unions. Rodrigo Nunes convincingly argued that political strategies aspiring to have a transformative impact must articulate these organisational dimensions in ways that are appropriate to each context.[6]

The cleavage underlying the distinction between the two transitions is rather, and unsurprisingly, that of labour versus capital. The ecological transition from above designates projects for a transition in the interest of capital, a transition that does not question the centrality of the profit imperative. Vice versa, the ecological transition from below is a project for the transformation of production geared towards environmental sustainability in the interest of the working class. (As noted in Chapter 2, any talk of interests is actually a political proposition in search of validation by a constituency.) The 'below', as I see it, simply refers to the fact that in the incumbent mode of production, capital rules 'above' the

5 Lorenzo Feltrin and Emanuele Leonardi, 'Working-Class Environmentalism and Climate Justice: The Challenge of Convergence Today' [2022], *Project PPPR*, 15 February 2023, projectpppr.org; Paola Imperatore and Emanuele Leonardi, *L'era della giustizia climatica: Prospettive politiche per una transizione ecologica dal basso*, Naples-Salerno: Orthotes, 2023.

6 Rodrigo Nunes, *Neither Vertical nor Horizontal: A Theory of Political Organization*, London: Verso, 2021.

working class. Therefore, any political project having a systemic alternative as its ultimate horizon must start from conflictual processes rooted 'below', through organisational forms that cannot be dogmatically predetermined in advance.

Why then not simply talk of a 'workers' ecological transition'? I propose that the phrase 'ecological transition from below' better points to an intersectional conception of class struggle, articulated with the fight against patriarchy, racism and other systemic hierarchies, all of which have their 'above' and their 'below'. And can we not resort to a 'just transition'? We can – after all, it is possible to define the concept of just transition along the lines I have just put forward. However, at the least in the Italian scene, we felt that – after the adoption of the just transition framework by the UN and other mainstream institutions – we needed a different phrase to convey our perspective.

In my view, three key distinctive albeit non-exhaustive features of the ecological transition from below can be highlighted: decommodification of production and nature, redistribution of wealth, and a rebalancing of the international division of labour.

First, decommodification refers to the production of wealth not for the market, but for the direct satisfaction of needs. Like the molecules pumped into petrochemical cracking plants, life in capitalism is cracked by the frontier of commodification breaking production apart from reproduction. Commodified production, based on dispossession and driven by profit, is structurally geared towards environmental degradation, such that a key element of future ecological transitions should involve pushing back the frontier of commodification to bring more wealth creation outside of the market. This is not merely a question of public and common ownership versus private property, as many state-owned enterprises and cooperatives sell their products on the market for profit. It is rather the difference, for instance, between free universal healthcare and private clinics providing treatment at market prices according to the ability to pay. While welfare services are, by definition, decommodified (because they are not sold at their market price), more goods and services could be provided as valueless wealth: food, energy, housing, transport, potentially everything.

Social form is not, so to speak, merely formal. Throughout this book, it has been argued that the commodity form shapes the material content of capitalist work and technology. Therefore, decommodification also opens up new possibilities for alternative technological systems and labour processes. The more the production of wealth is freed from even indirect subordination to the production of capital, the wider such possibilities are. Returning to the example of electric transport, while the commodity form is predisposed to a mobility system centred on private vehicles, decommodified production could prioritise an infrastructure centred around multi-modal public transport. Of the two, only the latter can ensure less consumption of raw materials, less waste and a more egalitarian access to mobility. However, capitalism can only tolerate a limited amount of wealth being produced outside of the commodity form. Therefore, the determination of the latter's boundaries is always an antagonistic process, carved out by the balance of class power.

Second, redistribution is about reducing inequality. Any transformation of production, whether actually 'ecological' or otherwise, should guarantee the dignified livelihoods of workers. Redistribution within capitalism is a good-in-itself, simply because it means a better life for more people. Ultimately, however, it is not about 'a fair day's wage for a fair day's work'. Rather, it is about changing the balance of class power between labour and capital by compressing the intra-working-class hierarchies standing in the way of its political recomposition.

On a macro level, redistribution can come about in different ways, such as inversely proportional wage increases or progressive tax reforms. The reduction of working hours with no pay cuts is an important form of redistribution, as the resulting liberated time is an opportunity to practise ways of life that are foreign to capitalist productivism and consumerism. A universal basic income can also contribute to delinking the reproduction of the working class from the production of capital. However, this can only be effective if it is not conceived merely in monetary terms, but also as the direct and unconditional provision of certain goods and services. For example, giving money to everyone so that they can buy health services from for-profit clinics, as in cash-for-care schemes, is still better than denying treatment to those who cannot afford it. Nonetheless, the direct provision of treatment outside of the cash nexus is a higher form of redistribution, as it liberates healthcare from direct commodification, making it more amenable to a prevention-centred approach and democratic planning outside of the profit imperative.

Third, rebalancing refers to moving towards a non-hierarchical international division of labour. While the West and China compete for technological supremacy, other countries and regions are relegated to providing cheap labour and cheap nature. As Leopoldina Fortunati aptly put it, 'The working class was born with the deep internal divisions of capitalist class relations and . . . the terrain of the class composition of the workers has itself always been a terrain of class struggle.'[7] If we are serious about struggling against the intra-working-class stratifications that obstruct international solidarity, then we must conclude that technological capabilities must be more evenly diffused globally, and industrial production in the Global North will have to degrow.

For example, the ex-GKN dispute is not about defending European industry per se, along with its colonial character and history. Rather, it is about the transformation of both production and the social relations that shape it. Specifically, in this case, it is about a transformation and decommodification of transport and energy that would require fewer raw materials in the first place. Ultimately, this means pushing against the scramble for critical minerals currently waged by the white plan of capital and, instead, strengthening international ties between the forces of the ecological transition from below.

An internationalist politics must articulate desertion from war together with struggles for social and environmental rights. Because the white plan of capital marked an abandonment of the ecological transition from above in the Global North, the ecological transition from below is currently the sole project for an ecological transition there. As the GKN Factory Collective stated: 'Rearmament is the explicit negation of any climate transition goal: the military sector and war are by

7 Leopoldina Fortunati, *The Arcana of Reproduction: Housewives, Prostitutes, Workers and Capital*, London: Verso, 2025 [1981], 233.

definition pollution . . . That is why today our economic plan is the only plan at stake to preserve life.'[8]

'Very well,' the sympathetic reader may think, 'but who are the social actors that can wage the struggle to advance this ecological transition from below? Who will build the tremendous amount of counterpower needed to gain meaningful concessions from state and capital and contain the disastrous damage that is already unfolding before our eyes?' In searching for answers, I suggest that we should look at interests and power within the global working-class composition. Those who have the strongest interest in limiting the ecological crisis are those most affected by it and who benefit the least from the *status quo*, like the peasants-in-limbo who have been losing their livelihoods to the ecological crisis. On the other side, those who have the greatest power in this context are the workers employed in the most environmentally destructive capital-intensive workplaces, like the employees of coal-fired power plants and oil refineries.

Coalition-building between these two types of working-class compositions would be needed to generate the broad social mobilisations necessary for working-class rigidity to become a more assertive counterpower to capitalist noxiousness. However, this cannot happen by the *fiat* of theoretical preaching. In fact, when it comes to the will and ability to challenge systemic parameters, interest and power tend to be inversely proportional. The objective fragmentation of the technical and social compositions of the working class means that *opportunities for convergence tend to arise in conjunctures of crisis*, that is, in situations when and where the structures separating different working-class segments are shaken. This is why opportunities must be seized quickly – before such structures ossify again. Indeed, the convergence between labour and climate justice struggles successfully led by the GKN Factory Collective only became possible *after* the factory closure was announced, but *before* the machinery was taken out.

The convergence between workplaces and communities also requires

8 Collettivo Di Fabbrica – Lavoratori Gkn Firenze, 'Abbiamo un piano. L'unico,' Facebook, 22 March 2025.

the previous sedimentation of dense organisational layers. As argued in the introduction, an element of *operaismo* still useful today is the principle that different working-class compositions require different organisational forms. Throughout this book, we have seen how the technical (workplace) and social (community) compositions of the working class are fragmented and stratified along multiple lines, being constantly reconfigured by cycles of struggle and restructuring. As a result, there is no universal formula that can be uniformly applied to all contexts; working-class rigidity requires organisational flexibility. Re-working the Trontian motto 'strategy to the class, tactics to the party', we could say 'rigidity to the class, flexibility to the organisation'.[9]

Under this realm of flexibility, which applies more generally to the organisational question, I propose putting the issue of state power. The limits of the capitalist state as a tool for socialist transitioning have been convincingly expounded already. Simply put, the capitalist state depends on capital accumulation for its own funding, therefore it cannot be expected to put such accumulation in jeopardy.[10] However, for the purposes of wrestling particular wins along the lines of decommodification, redistribution and rebalancing, the questions of who holds state power and how are not a matter of indifference.

For core workers in large-scale industry and, more generally, for workers in capital-intensive sectors, large unions are – in normal times and despite all their ambiguities – an effective means of defending their interests and negotiating relatively good wages and conditions through enforceable collective agreements. Large unions also tend to have considerable influence in the public sector. However, the further one moves towards the precarious pole of the working class, the harder it becomes for the big unions to organise a highly fragmented and dispersed workforce, as there is little chance of clogging the bottlenecks of production from within the workplace. A multiplicity of organisational forms is thus adopted by different segments of the surplus working class: smaller and more radical unions, social movement organisations, community

9 Mario Tronti, *Workers and Capital*, London: Verso, 2019 [1966].

10 Simon Clarke (ed.), *The State Debate*, London: Palgrave Macmillan, 1991.

associations, informal committees in charge of enforcing roadblocks, etc. The party form, itself a very broad category with multiple possibilities, is present in all segments of the working class, although today it does not have the same prominence as in other eras.

Because precarious workers face serious obstacles in organising within the workplace, as a result of their employment insecurity and fragmentation, many of their struggles are community-centred. They have the urban neighbourhood or the rural village as their base and are played out as roadblocks, demonstrations, and riots in the public space.[11] It is true, as Silvia Federici wrote, that capital prefers 'cooperation at the point of production, separation and atomization at the point of reproduction'.[12] However, in addition to the home, there are collective sites of reproduction such as educational institutions, public squares, cultural centres and sports facilities. These often serve as the infrastructure, as well as the object, of community-centred mobilisations. For example, the 2013 movement against Turkish PM Recep Tayyip Erdoğan's authoritarian turn began in defence of a public park. In community-centred environmental mobilisations too, reproductive settings often become the base for organisation, while the expansion of welfare services is a frequent demand. For instance, Quintero and Puchuncaví's high schools were an important centre of community organising against pollution, while the reinforcement of local public healthcare was one of the movement's requests.

As previously argued, the community – as the sphere of workers-as-reproduced – is the realm in which workers more directly experience the material interest to push back against the ecological crisis. The challenge for working-class environmentalism today is to generate a political recomposition and convergence of workplace and community struggles through the elaboration of common platforms geared towards breaking 'job blackmail'. This challenge is therefore also one of convergence

11 Ana C. Dinerstein, *The Politics of Autonomy in Latin America: The Art of Organising Hope*, London: Palgrave Macmillan, 2015.

12 Silvia Federici, *Revolution at Point Zero: Housework, Reproduction, and Feminist Struggle*, Oakland, CA: PM Press, 2012, 146.

between different forms of organisation, where convergence does not mean merging or becoming similar. Rather, it means cooperation between different organisations from their base, while maintaining the specificities that allow them to function effectively for their respective working-class segments.

A conceivable sequence is that unrest begins with the working-class segments who have more interest in change, that is, the most precarious ones, especially in the Global South. Their agitation is mainly waged in the communities, where they have better chances to organise. The mobilisations of precarious workers exert a pressure on the more secure ones, who are sometimes moved to bring the unrest into the workplaces. The political recomposition of precarious and secure, community and workplace, in turn forces state and capital to make concessions. This was, very broadly speaking, the pattern we have seen in several uprisings of the past years, from the Tunisian Revolution to the Chilean *estallido social*.[13] The trouble is that working-class decomposition is then enforced through the crises and restructurings imposed by international competition, which acts as a disciplinary force eroding to varying extents the gains won by the mobilisations. Sometimes, such downturns combine with state repression, imperialist interventions and domestic destabilising forces to exacerbate intra-working-class conflicts, even reaching all-out civil war. Decommodification is crucial as a shield for working-class counterpower against the law of value, but a denser working-class organisational ecology is needed to reinforce such a protection.

As seen throughout this book, life in capitalism is separated by the line of value. It is cracked by the frontier of commodification between production and reproduction, with the latter subordinated to the former. The conditionality of workers' reproduction to capitalist work is the material basis of working-class denialism (that is, the diffusion of far right climate denialism within segments of the working class), because

13 Lorenzo Feltrin, 'Between the Hammer and the Anvil: The Trade Unions and the 2011 Arab Uprisings in Morocco and Tunisia', PhD thesis, University of Warwick, 2018; Sebastián Osorio and Diego Velásquez, 'El poder sindical en el "estallido social" chileno: La huelga general de noviembre de 2019', *Revista española de sociología* 31(1), 2022, 1–21.

capitalist work depends in turn on the ever-expanding production of capital, and thus on the accumulation of ecological crisis. In contrast, reproductive interests are the material basis of working-class environmentalism, because the noxiousness inherent to capital accumulation imperils the workers' own ecological conditions of reproduction. Going back to our point of departure, the working class in the ecological crisis really counts twice: one time as a force that *produces* the ecological crisis, and another time as a force that *refuses* to produce it; one time *in* the ecological crisis, another time *against* it. The challenge is that of building convergences between workplace and community organising to turn working-class environmentalism into a winning alternative to working-class denialism, creating the conditions to reach the end of the month while moving beyond the end of the world.